W9-CSP-640

progress in
filtration and
separation

progress in filtration and separation

edited by

r.j. wakeman

Department of Chemical Engineering,
University of Exeter, Exeter (Great Britain)

ELSEVIER SCIENTIFIC PUBLISHING COMPANY

Amsterdam — Oxford — New York   1981

6387 - 9943
Chemistry Lib.

XD82
1413
CHEM

ELSEVIER SCIENTIFIC PUBLISHING COMPANY
Molenwerf 1
P.O. Box 211, 1000 AE Amsterdam, The Netherlands

Distributors for the United States and Canada:

ELSEVIER/NORTH-HOLLAND INC.
52, Vanderbilt Avenue
New York, N.Y. 10017

ISBN 0-444-42006-1 (Vol. 2)
ISBN 0-444-41820-2 (Series)

© Elsevier Scientific Publishing Company, 1981
All rights reserved. No part of this publication may be reproduced, stored in a retrieval system or transmitted in any form or by any means, electronic, mechanical, photocopying, recording or other-wise, without the prior written permission of the publisher, Elsevier Scientific Publishing Company, P.O. Box 330, 1000 AH Amsterdam, The Netherlands

Printed in The Netherlands

TP156
F6P7
V.2
CHEM

v

INTRODUCTION

Whenever the subject of filtration and separation is
aired it is invariably only a short time before one hears the
adage that these unit operations are more of an art than a
science. Whether or not this cliche was coined by practitioners
of the art who preferred empiricism to theoretical approaches
is a moot point, but it is a fact that practising engineers
know very little of the scientific background of filtration,
or of the attempts to provide more substantial design procedures
in the filtration and separation field.

Since the 1st World Filtration Congress held in Paris
in 1974 there has been a noticeable increase in the number of
academic and industrial researchers taking an active interest
in filtration and separation subjects, making significant contributions to the science and technology of solid-fluid separation processes. The importance of particle science, technology and separations, in for example, the chemical, food and
allied industries, mineral and solid fuel processing, the
protection of the environment from the wastes of communities
and industries, is perhaps becoming recognised as a key area
where fundamental and applied research may pay good dividends.
Improved understanding between theory and practice, and the
development of better models and in some cases superior equipment design, should emerge from this greater research input.

The aims of this multi-author book series remain the
same as with the previous volume; to provide researchers and
engineers with critical reviews and carefully considered opinions
of recent works in the field of filtration and separation and
adjoining areas. Whilst some chapters are intended to bring
older topics up to date, others gather and examine the results
of new or freshly utilized techniques and methods of seeming
promise. As before, the authors have furnished definitive
reports giving their own analyses of the subjects, not merely
annotated bibliographies. Individual authors have exercised
their judgement as to which are significant papers in their own
areas of expertise, and all have contributed significantly to

the research about which they write. Perhaps this series will
play its part in helping to put filtration and separation on
to a more scientific basis; and perhaps the elements of art
and faith in the design of separation equipment will be
substantially reduced in the future.

Richard J. Wakeman

Department of Chemical Engineering
University of Exeter,
Exeter, Devon, U.K.

CONTENTS

MECHANISMS OF DUST COLLECTION IN CYCLONES

JOHN ABRAHAMSON

Chemical Engineering Department, University of Canterbury, Christchurch, New Zealand.

ABSTRACT

In this discussion of cyclone dust collectors, traditional design criteria are first compared with experimental experience, and are found inadequate. Following this an outline is given of the importance of vortex breakdown, precessing cores and wall vortices. A quantitative evaluation is then given as far as is possible of different physical factors in dust collection, both in the approach duct and in the cyclone; dispersion, agglomeration, centrifugal collection, turbulent diffusion, and electrical effects. It is suggested that four dimensionless numbers are needed to characterise cyclone operation for coarse dust, together with an electrical charging number for fine dust. Finally some newer cyclone variants are assessed.

CONTENTS

1. INTRODUCTION

1.1 Design procedures

The cyclone dust collector is one of the simplest devices to construct, and is often the lowest cost alternative for collection of dust larger than 5 μm. It is this simplicity which allows process engineers commonly to design their own cyclone for a particular duty. In most instances, a detailed knowledge of the dust to be separated is not available, so that a complete estimation of the efficiency is not appropriate. In such cases, the engineer selects a *geometry* of cyclone which has for others achieved adequate collection for a reasonable pressure drop, and scales the size (and number of cyclones) to his process conditions. Typically he will then put his cyclone into use without ever measuring its collection efficiency. He will rely on his having used perhaps a "best practical means" of dust removal, and will rely on his exhaust measurements to ensure that emission is below a certain limit.

In this chapter most space will be devoted to "reverse flow" cyclones, as these are most widely used and investigated, and conclusions about the physical mechanisms operating in these cyclones may still be helpful in understanding other types. Before launching into a discussion of mechanisms however it is helpful to use as a reference the currently "safest" design practice an engineer might use(at least if he was educated from European publications).

The geometries he might choose are shown in Fig. 1, from the best of English and Dutch experience (refs. 1, 2), for high efficiency, higher pressure cyclones (A, B) and for high-throughput lower pressure cyclones (C, D) with lower collection efficiency. The performance may be summarised adequately for him by the numerical value of two dimensionless numbers;

(a) a Stokes number

$$Stk_{\eta} = \frac{d^2_{p(\eta)} V_i \rho_p}{\mu D} \qquad (1)$$

corresponding to the particle diameter $d_{p\eta}$ for which say $\eta = 50$ or 80 percent of particles are collected (V_i is the velocity in the inlet of a cyclone of barrel diameter D, ρ_p is the particle density and μ is the gas viscosity).

(b) the number of inlet velocity heads in the cyclone pressure drop ΔP

$$\zeta = \frac{2\Delta P}{\rho V_i^2} \qquad (2)$$

(here ρ is the gas density)

Our engineer can calculate the cyclone sizes (from the geometries of Fig. 1 scaled on D) for a desired particle size above which most dust is collected (d_{p50} or d_{p80}), since he knows from published experiments that Stk_{50} (and Stk_{80}) has a particular value for each cyclone type. He can also estimate the pressure drop knowing ζ , which again is dependent only on the cyclone geometry. The values of these two parameters are shown in Table 1 (the square root of Stk_{50} is listed, as it is used as a dimensionless form of particle diameter)

TABLE 1
Values of dimensionless "cut" particle diameters and pressure drops for Stairmand and van Ebbenhorst Tengbergen cyclones

Cyclone type (as for Fig. 1)	Ref.	$\sqrt{Stk_{50}}$	$\sqrt{Stk_{80}}$	ζ	$\sqrt{Stk_{50}}\ h/2r_o$
A	1	0.18	0.27	5.0	0.50
B	2	< 0.04[a]	0.08[a]	25[b]	< 0.1
C	1	0.50	1.5	6.5	1.0
D	2	0.18[a]	0.55[a]	10.5[b]	0.4

[a] Reported for a dust concentration of 10 g m^{-3}
[b] Values for clean air; ζ drops to 19.5, 8 at high dust concentrations.

It can be seen for a given size of cyclone and inlet velocity (commonly taken around 15 ms^{-1}), type B outperforms type A (the Stairmand high efficiency cyclone), at the cost of a much higher pressure drop. To choose the cyclone most suitable for cleaning a given flow of gas Q, the engineer can substitute for V_i in terms of Q and the inlet area in eqns. 1 and 2, perhaps restricting the range of V_i to 10-30 m s^{-1} for "adequate" operation. If he wants to make a comparison between different types of cyclones using economic parameters, he can use dimensionless power consumption and dust removal factors, as Kalmykov et. al. (ref. 3) have done for various Soviet and European cyclones. If he has a complete size distribution of the dust to be collected, he can calculate an overall collection efficiency, by taking not only Stk_{50} and Stk_{80} as in Table 1, but Stk for several other efficiencies ζ , and so plot a $\zeta - d_p$ curve. Multiplication of ζ with mass fractions corresponding to the same d_p size interval, and summing, yields the overall efficiency.

The simplicity of this design approach is rather deceptive, as is also the simplicity of the theory that lies behind it. The list of assumptions usually made supporting the use of the Stokes number and the pressure coefficient may be given as;

1. In all geometrically similar cyclones the gas velocities at all geometrically similar points are strictly proportional to the inlet velocity.

2. The dust is collected as if it was in a laminar flow, whereas the pressure drop is calculated for turbulent flow.

3. Dust particles are treated in isolation - there is no interaction between them, and the gas flow is not modified by their presence.

4. Particles migrate to the cyclone wall under a centrifugal force opposing a viscous drag force, and are immediately and finally collected at the wall (transport to the bin is wholly efficient).

5. The only forces operative are steady viscous and inertial forces, calculated for spherical dust particles.

In the theory, attention is most often focussed on a "critical collecting zone" just below the gas exit tube, where the swirl velocity is found to be high. A radial inward drift of gas occurs, (a "vortex sink" flow with a velocity much less than the swirl velocity) taking particles in with it so that they are lost to the exhaust. If a given particle is to be collected in the cyclone, the centrifugal force given it by travelling at the gas swirl velocity V_θ, will overcome the inward drag force due to the radial gas velocity V_r. Where the two forces balance, the particle remains at the same radius r, i.e.

$$\frac{\pi d_p^3}{6} \, \rho_p \, \frac{V_\theta^2}{r} = 3\pi\mu \, d_p \, V_r \qquad (3)$$

using Stokes's law for a spherical particle. Thus the following groups may be equated

$$\frac{d^2_p \, V_\theta \, \rho_p}{\mu \, r} = 18 \frac{V_r}{V_\theta} \tag{4}$$

for a particle which has equal chance of being captured or lost (due to random perturbations). It can be seen that using assumption 1 above, V_r/V_θ is a specific value for a given design, (the critical collecting zone is at a specific location) and V_θ, r are specific fractions of V_i, D respectively. Substituting, we obtain equation 1 where the Stokes number is that value appropriate for 50% collection. It is not so obvious that this argument can be extended to other efficiencies, as the relative radial velocity to be used in eqn. 3 is then $V_r - V_{rp}$ where V_{rp} is the radial particle velocity. The Stokes number value will then depend on the particle properties as well as that of the cyclone.

Many attempts have been made to extend the critical collecting zone over the length of the cyclone, thereby bringing the geometry into the argument. Thus Barth (ref. 4) considers a cylindrical collecting surface of the same radius as the gas exhaust, extending down to the bottom of the cyclone, with gas travelling down towards the bin on the outside, and gas travelling up to the exhaust on the inside. He makes two assumptions; that all particles cross this collecting surface only once, and that the total gas flow is uniformly distributed over the collecting surface, hence giving an easy estimate of radial gas velocity. With the inclusion of a radial dependence of swirl velocity

$$V_\theta \propto 1/r^m \tag{5}$$

where m varies from 0.5 to 0.7 by experiment, the particle size for 50 per cent collection is (ref. 4)

$$d_{p50} = C_{50} \left(\frac{\mu}{\rho_p h} \right)^{\frac{1}{2}} R_e^m \tag{6}$$

where h is the height of the cyclone from exhaust tube to dust bin, R_e is the radius of the gas exhaust, and C_{50} is a constant to be determined by experiment. This treatment predicts correctly that the "cut" particle size d_{p50} will diminish with increasing length and smaller exhaust openings, within limits. In so doing, it parallels some earlier treatments of cyclone geometry in terms of the number of turns that the gas makes in travelling down the outer wall of the cyclone. In these schemes, the longer the gas path, the more likely it is that a certain particle will be captured.

An apparently different approach to the prediction of cyclone efficiencies has been made by Leith and Licht (ref. 5), who assume that dust is instaneously redistributed across any section normal to the cyclone axis. However, they have merely enlarged the collecting surface until it coincides with the inside wall, and the "turbulent mixing" within the cyclone volume redistributes the dust remaining after laminar-type migration to the wall. It is not surprising that they also find their predicted efficiency to be a function of a Stokes number, and a dimensionless group related to cyclone geometry.

It can be seen that the use of a Stokes number is not inconsistent with some turbulent diffusion or mixing of particles within a cyclone, even before the particles reach the separating zone. No account is kept of the none-to-one trajectories of gas or particle from the inlet to the collecting zone in the Stokes number theories, in contrast to the "number of turns" early work, exemplified by Lapple (ref. 6).

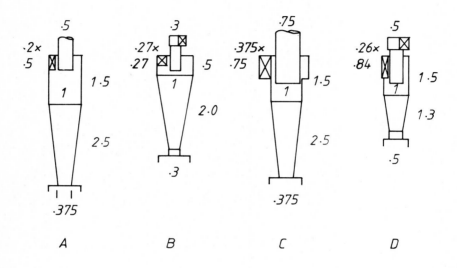

Fig. 1. Geometries of Stairmand and van Ebbenhorst Tengbergen cyclones, relative ro the barrel diameter.

1.2   Experimental tests of the design theories

The Stokes' number scaling methods appear to apply over limited ranges of the physical factors; Tengbergen (ref. 2) has published results for cyclones with diameters 280 to 620 mm and for two dusts, clay and foundry dust. Stairmand does not report supporting results at all but recommends from his experience that "large transpositions of powder grading, gas viscosity, powder density and throughput should not be made". A 2 to 1 ratio of gas velocity was acceptable.

The effect of raising the gas temperature has been found in practice to diminish cyclone collection efficiencies (refs. 6, 7), as expected from the effect of a

higher viscosity on the Stokes number. However, the only quantitative measure of
the effect on the Stokes number in an industrial cyclone appears to be that from
Whiton (ref. 7) who measured the collection of fly ash with a 610 mm diameter
cyclone. For air from 300 K → 530 K, the d_{p90} rose 65 → 150 μm (constant V_i) compared
to the constant Stk prediction 65 → 80 μm, which is much smaller.

    This stronger dependence on viscosity is supported by Smith, Wilson and Harris's
(ref. 8) study with small sampling cyclones. They found for cyclones of 15-45 mm
diameter and air temperatures from 300 K to 480 K that d_{p50} increased <u>linearly</u> with
viscosity (in this author's opinion, proportional to within likely error - see
Fig. 2). The effect of higher gas pressure is to lower cyclone efficiency, at a

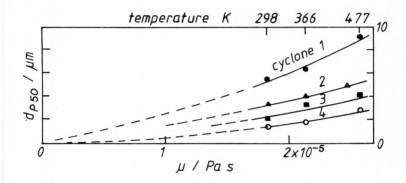

Fig. 2 Cut diameters with different temperatures (viscosities) for four sampling
cyclones (ref. 8).

constant velocity. Knowlton and Bachovchin (ref. 9) show in their Fig. 7 that
d_{p90} increases from 4 to 6 μm for an increase in air pressure from 3 bar to 55
bar, with high solids loading (4025 g m^{-3}) and inlet velocity (25 m s^{-1}). The
increase in air viscosity is only about 13 percent for this pressure increase,
so even a proportional increase in d_p would not account for the experimental
increase. It is more likely that the gas density is important. This increases
17 fold in Knowlton and Bachovchin's experiments, indicating a $\rho^{0.14}$ dependence
for d_p.

    If we accept a gas density effect at higher pressures, we must also accept it
at higher temperatures. However a $\rho^{0.14}$ dependance yields only a 7 percent rise in
d_p for a temperature rise from 298 to 477 K, compared to Smith, Wilson and Harris's
(ref. 8) experimental average of 70 per cent increase in d_p. If gas density and
viscosity are the only factors, viscosity must be much more important, and approximately d_p is proportional to μ.

Changes of $d_{p\eta}$ associated with a change in particle density ρ_p are also little investigated. Stern et al (ref. 10) describe Dennis et al's experiments with dust of ρ_p = 2700 to 7800 kg m^{-3}, from 0.5 to 8 μm in a 51 mm diameter cyclone, and it is clear that while $d_{p\eta}$ diminished with increase in ρ_p, it diminished inversely to ρ_p rather than to $\rho_p^{\frac{1}{2}}$ for ρ_p ≃ 2700 to 4000 kg m^{-3}. At higher densities the dependence was closer to $\rho_p^{\frac{1}{2}}$. Smith, Wilson and Harris (ref. 8) however, also working with small cyclones, found evidence for a $\rho_p^{\frac{1}{2}}$ inverse relationship in the low density range (1000-2000 kg m^{-3}) as did Cushing et al, cited by Smith, Wilson and Harris. Presumably Tengbergen (ref. 2) also found a $\rho_p^{\frac{1}{2}}$ inverse dependence with his larger cyclones.

Changes in cyclone velocity have been dominant in theories but again poorly investigated in relation to the Stk prediction. It is recognised that efficiency rises steeply until the inlet velocity reaches 5-10 m s^{-1}, and rises gradually above that until a plateau or even a well defined maximum is reached at 25-50 m s^{-1} (e.g. ref. 11). In the gradually rising regime, changes in d_{p50} with velocity have been reported for small sampling cyclones and one larger cyclone. Chan and Lippmann (ref. 12) find that compared to the predicted inverse $v^{0.5}$ dependence from Stk = constant, inverse $v^{0.75}$ to $v^{1.29}$ covered the range of most of their cyclones. They reported one larger cyclone (diameter 270 mm) with an inverse $v^{0.98}$ relationship. The smallest cyclone studied (10 mm diameter nylon) showed two definite extreme dependencies, with a shart transition at a certain velocity. Smith, Wilson and Harris (ref. 8) also show inverse $v^{0.63}$ to $v^{1.11}$ behaviour for d_{p50}, for five small cyclones.

Finally much interest is attached to scaling up a cyclone by its characteristic dimension D. The sole importance of Stk in this respect has been recently challenged by Beeckmans and Kim (ref. 13), who studied two geometrically similar cyclones 76 and 152 mm diameter with a range of velocities. Their careful work included precautions such as neutralisation of electrical charges and control of humidity. They showed that the inclusion of Reynolds number

$$Re = \frac{D\,V_i\,\mu}{\rho} \tag{7}$$

in the correlation significantly improved the fit. These conclusions also applied to their analysis of small sampling cyclone measurements reported in ref. 12. However, these analyses are limited to Re < 10^5, and industrially interesting scale-up corresponds to Re significantly above this. Comparison of efficiencies from cyclones differing widely in size, is expected to give a more sensitive indication of the Re dependance. Zverev and Ushakov (ref. 14) show a plot of efficiency versus Stk Re$^{0.5}$ $(\rho/\rho_p)^{0.5}$ for a series of Ts N - 15 Soviet cyclones with diameters 100 to 800 mm. The experimental points lay adequately close to a single curve for this long bodied cyclone (length 4 barrel diameters). However Reznik, Prokofichev and Matsnev (ref. 15) have tested this criterion more severely by comparing Ts N - 15 cyclones for D = 500 and 3750 mm, and found widely differing curves for the different sizes. They also found this discrepancy with a more squat cyclone over a similar size range.

In finding a relationship Stk Re^a which brings Reznik, Prokofichev and Matsnev's points onto the same curve, one must point out what appears to be a mistake in labelling of their curves in their Fig. 5. As labelled, their results show a smaller d_{p50} for the larger cyclones, which is unlikely. If one reverses the labels (small D for large D) their results for the smaller cyclone lie much closer to Zverev and Ushakov's curve for smaller cyclones, and a larger d_{p50} is shown for larger D. In this case the efficiency is correlated as a unique function of Stk $r_e^{-0.25}$ for the different D. Matsnev and Ushakov (ref. 16) have shown that a Stk versus efficiency plot is clearly inadequate to describe both a 500 mm and a 2600 mm "uniflow" or straight-through cyclone, especially at low particle sizes. They have found that an adequate group is Stk $Re^{-0.1}$, for the range $1.8 \times 10^5 <$ $R_e < 10^6$ and $0.05 < Stk < 1.5$.

From the above experimental studies, it would appear that a group

$$\frac{d_p \, \rho_p^{\frac{1}{2}} v_i}{\mu \, D^{2/3} \rho^{0.1}} \; = \; G \qquad\qquad (8)$$

although dimensional, might be a practical compromise for industrial cyclone scaling.

There are however, some factors not yet covered, which are important even at this crude predictive level. One is the influence of particle shape and state of dispersion, which are allowed for by taking the "aerodynamic diameter" for the d_p to be used in the scaling group. This aerodynamic diameter is the diameter of the sphere with the same density as the primary particles, ρ_p, which would have the same terminal velocity under gravity. Not all the literature quoted above uses this refinement, which involves *an adequate* dispersion of the dust, and some classification on an instrument measuring falling velocities. Doerschlag and Miczek (ref. 17) show the marked difference in efficiency-particle size curves which can occur with different methods of dispersal and classification.

Another factor is the concentration of particles. Tengbergen (ref. 2) has given the concentration dependence for his conditions explicitly, by drawing a different Stk-efficiency curve for each concentration. The efficiency generally increases with increase in dust concentration, and Tengbergen shows this effect

is most dramatic for the smaller particle sizes, and for these sizes can be larger
than any normal scaling corrections. In some instances the efficiency has been
reported as diminishing with increase in concentration (e.g. refs. 9, 10).

Finally in this section some comments will be made on the effect of changing
the *geometry* of a cyclone. Purely empirical investigations have been made by
Ter Linden (ref. 18) and Cherrett (ref. 19) on the effect of changing inlet
area, cone length and gas outlet diameter while keeping the barrel diameter
constant. These efforts have been useful in establishing optimal geometries
for a given basic size of cyclone, but use of the results may be risky when an
extrapolation to other sizes and conditions is required. Considerable effort
has been made to construct geometry-related theories, particularly by German authors,
and in assessing these we will first limit our observations to the four geometries
shown in Fig. 1, and to Barth's prediction eqn. 6. Taking the swirl velocity-
radius exponent m as 0.5, eqn. (6) gives d_{p50} as inversely proportional to
$(h/2r_o)^{\frac{1}{2}}$ or the square root of the number of exhaust diameters separating the
gas and dust exits, if all other factors are constant. As a consequence, $d_{p50} \times$
$(h/2r_o)^{\frac{1}{2}}$ should be constant for all geometries if all other factors are constant
i.e. $\sqrt{Stk_{50} \ h/2r_o}$ should be constant. This latter group has been listed for
the four cyclones of Fig. 1, in Table 1. It can be seen that its value varies
considerably, putting in some doubt the adequacy of eqn. 6 to allow for different
geometries.

Even more ambitious are attempts to predict complete efficiency-particle size
curves from the geometry and conditions of the cyclone. Leith and Licht (ref. 5)
have compared various theories with the experimental data of Stairmand (ref. 1)
(203 mm dia. cyclone, silica),Peterson and Whitby (ref. 20) (305 mm dia. monodisperse ρ_p = 1600 kg m^{-3}, and Tengbergen (ref. 2). They found their own theory
fitted the data to within 10 per cent or better, but other theories (Barth,
Lapple, Sproull) compared only poorly (at least with Stairmand's data).
Rausch (ref.21) compared experimental results from several German sources with
Barth's theory, finding a large discrepancy, and with his own theory, a potential
flow calculation of flow just in front of the gas exit, showing good agreement at
least in the shape of the curve. Avant, Parnell and Sorenson (ref. 22) collected
sorghum dust with 9 cyclones of different shapes, and could find no satisfactory
agreement with either Muschelknautz's theory (an extension of Barth's model) or
Leith and Licht's theory. Ayer and Hochstrasser (ref. 23) on the other hand,
found good agreement between Barth's theory and experimental efficiency-particle
size curves when using a Stairmand "high efficiency" (Fig. 1 type A) cyclone of
only 19 mm dia. Leigh and Licht's theory performed poorly with the small
Stairmand cyclone, but as Hochstrasser (ref. 22) points out, their theory predicts
experimental efficiency curves well for moderate sized industrial scale Stairmand

cyclones, with data collected from a number of sources (Stairmand ref. 1, Van Schaick ref. 23, and Cherrett ref. 24).

---

2. VORTEX BEHAVIOUR AND CYCLONE PRESSURE DROP

In this section an account will be given of recent observations of vortex behaviour, and these related, in some respects for the first time, to gas flow in cyclone dust collectors. Some more fundamental reasons will be seen for some of the standard designs, and perhaps a clearer idea obtained for the pressure drop mechanism in a cyclone. The phenomena of *vortex breakdown* is often important in the trailing vortex behind an aeroplane's wing, and indeed appears to be the mechanism of production of eddies of smaller scale in turbulent flow away from a boundary (ref. 25). Vortex breakdowns will also be seen to be dominating cyclone flows.

We can understand vortex behaviour in industrial equipment more clearly if we first consider some unconfined swirling flows, and introduce some useful concepts and results through these mathematically more tractable examples. In what follows, we will consider the motion of an incompressible fluid (density ρ = constant) of viscosity μ or kinematic viscosity $\nu = \mu/\rho$ within cylindrical coordinates r, θ, z (unless otherwise stated).

2.1 Streamline Vortices

The simplest unconfined flow is the "irrotational" potential flow vortex, independent of axial position z, where conservation of angular momentum and no rotation of individual elements of fluid (possible only for a flow where viscous effects are negligible) yield

$$V_\theta = c_\theta/r \tag{9}$$

The product rV_θ is a constant (c_θ) for this flow. This can be stated in another way by saying that the "circulation" line integral $\Gamma = 2\pi rV_\theta$ remains the same for all radii. Eqn. 9 becomes

$$\Gamma/\Gamma_o = 1 \tag{10}$$

where Γ_o is the circulation level which characterises the vortex, as given in Fig. 3(a).

Next we consider an unconfined flow more similar to cyclone flow; we consider a vortex with superimposed radial inflow and axial flow. Fluid moving in towards the axis is deccelerated and turns to accelerate in the axial direction. Burgers (ref. 26) has solved the cylindrically symmetric angular momentum equation

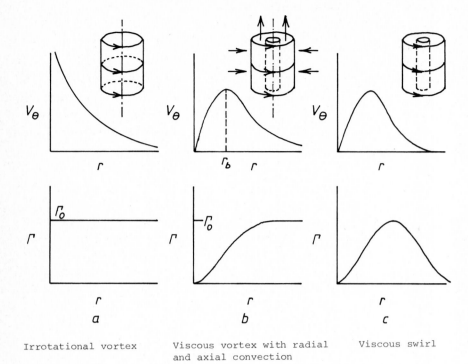

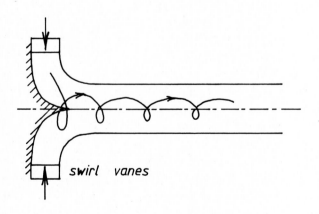

$$
\begin{array}{ccc}
a & b & c \\
\text{Irrotational vortex} & \begin{array}{c}\text{Viscous vortex with radial}\\\text{and axial convection}\end{array} & \text{Viscous swirl}
\end{array}
$$

Fig. 3. Various rotating flows

Fig. 4. Vortex generating duct

$$\frac{\partial V_\theta}{\partial t} + V_r \frac{\partial V_\theta}{\partial r} + \frac{V_r V_\theta}{r} = \nu \frac{\partial}{\partial r} (\frac{1}{r} \frac{\partial}{\partial r} (rV_\theta)) \tag{11}$$

for the special instance (A is constant)

$$V_r = -Ar, \; V_\theta = V_\theta(t, \; r), \; V_z = 2Az \tag{12}$$

satisfying the mass conservation equation, and momentum equations in the r, z
directions. The solution is independent of time, and is a variation on the
irrotational vortex form, in that

$$\frac{\Gamma}{\Gamma_o} = 1 - \exp\left[\frac{-Ar^2}{2\nu}\right] \tag{13}$$

This solution shows the inverse r relationship for V_θ at large r, together with a
maximum, and a steeply dropping V_θ for $r \to 0$, as shown in fig. 3(b). Thus with
one function we can model both the characteristic inner core of cyclone flow, and
its outer vortex, which have been often described as a "forced vortex" and a "free
vortex" respectively (the latter is empirically described by eqn. 5). The characteristic dimension is r_b at which the swirl velocity V_θ is a maximum, and $Ar_b^2/2\nu$
$= V_{rb} r_b/2\nu$ has the form of a radial Reynolds number. The Reynolds number
normally taken to describe a **viscous** vortex is the angular $Re_\theta = \Gamma_o/\nu$.

Uberoi (ref. 27) has pointed out the need for distinction between a *vortex*, in
which angular momentum can be convected in from a large radius ($\Gamma \to \Gamma_o$ for $r \to \infty$),
and a *swirl*, in which angular momentum is generated in a localised radial region
($\Gamma \to 0$ for $r \to \infty$). The difference between a vortex and a swirl is emphasised in
fig. 3(b) and (c). In practice we can anticipate that with cyclones, tangential
entry of fluid, especially multiple tangential entry, will approximate better to
the vortex form of velocity distribution within the cyclone barrel, whereas axial
entry through swirling vanes will approximate to the swirl form. Uberoi (ref. 27)
has mentioned the superposition of a swirl on a vortex.

It is important to emphasise the mechanism causing the drop of circulation and
velocity towards the axis in Burger's vortex. There is a loss of angular momentum
from any circular cross section of the vortex to the surroundings by molecular diffusion (viscous shear) for a laminar vortex. This opposes the inward convection of
momentum strongly towards the axis where the shear rates are high until little
angular momentum is available to be convected further towards the axis, and the core
rotates almost as a solid body (at the maximum angular velocity). In practical
experiments this core can be swelled by non-rotating fluid being fed axially
into the rotating flow. Harvey (ref. 28) generated a laminar vortex in an air
flow with the apparatus shown schematically in fig. 4. He measured angles of flow
in the tube consistent with the exponential form of equation 13, but found that the

size of the vortex core could be significantly reduced by removing (by suction applied through an annular slot), some of the boundary layer being shed from the centre point. He concluded that the viscous core was derived from this boundary layer.

More recently Faler and Leibovich (ref. 29) measured V_θ in water flowing through a similar vortex generator, and found good correspondence with the form of eqn. 13, i.e.

$$V_\theta = Kr^{-1} (1 - \exp(-\alpha r^2))$$ (14)

where K, α are constants. This correspondence failed in the region within 0.2 tube radius from the tube wall. Instead of having the uniform axial velocity V_z assumed in Burger's vortex, however, Faler and Leibovich found a faster jet in the centre, with a velocity distribution of the form

$$V_z = V_{z1} + V_{z2} \exp(-\alpha r^2)$$ (15)

with α common to both expressions. They have explained the faster jet simply from the larger axial pressure gradient at the centre. Most of their experiments were done with an axial pipe Reynolds number

$$Re_z = D\overline{V}_z/\nu$$ (16)

in the range 1000 to 8000 (laminar flow) (D is the pipe diameter, $\overline{V}_z$ is the average axial velocity. In this range, α was an increasing function of Re_z, and only weakly dependent on the strength of the angular momentum. Further experiments for Re_z up to 21000 (still laminar) apparently were also represented well by eqn. (14). Faler and Leibovich characterised their flows by two dimensionless groups - the Reynolds number Re_z and a dimensionless circulation number Ω, defined as

$$\Omega = \Gamma_{vanes}/\overline{V}_z D$$ (17)

where $\Gamma_{vanes} = 2\pi R_{vane} V_{\theta vane}$ and R_{vane} is the radius of the trailing edge of the swirl vanes, and $V_{\theta\,vane}$ is V_θ at that position. Assuming approximately constant Γ from the vortex-generating vanes in the tube wall, $\Gamma_{vanes} \simeq \Gamma_{wall}$, and

$$\Omega \simeq 2\pi R_{wall} V_{\theta wall}/\overline{V}_z D = \pi \frac{V_{\theta\,wall}}{\overline{V}_z}$$ (18)

2.2 Turbulent mechanisms in the core

A swirling flow is stable to disturbances, according to Rayleigh (ref. 30), if

$$\frac{d\Gamma^2}{dr} > 0 \qquad\qquad\qquad (19)$$

If $d\Gamma^2/dr$ is zero, as in the irrotational vortex, neutral stability exists.
Perhaps a more understandable statement of this criterion is that a non-viscous
fluid is unstable whenever the direction of local rotation (the vorticity) is
opposite to the direction of the overall rotation (the angular velocity). In
practice, flows approximating the irrotational vortex are stable to small disturbances, and Uberoi (ref. 27) gives evidence for this, and claims that a *vortex*
is stable no matter how large the Reynolds number Γ/ν becomes. On the other hand,
a *swirl* , as defined by Uberoi, is unstable because Γ decreases with increase in
r, and turbulence will appear.

It is important to note that Rayleigh's criterion 19 applies only for systems
with *one* velocity component $V_\theta(r, t)$. For flows with large relative axial velocities,
turbulence can be generated as in common axial pipe flow, and Uberoi claims that
the turbulence observed within trailing vortices behind aircraft wings comes from
the *axial* relative motion, and not from the rotational motion. This turbulence
appears to dampen out under the influence of the vortex when the axial velocities
are diminished. Ludweig has extended Rayleigh's analysis (ref. 31) to swirling flows
with radial gradients of axial velocity. He has shown theoretically and experimentally that eqn. 19 still applies, but that flows approximating a rigid rotation
are unstable to small dV_z/dr. He observed spiral disturbance vortices between two
coaxial rotating cylinders.

In industrial cyclones, however, little concern has been shown by designers to
achieve vortex rather than swirl conditions in the initial generation of rotating
motion (see however 5.2 and 5.3 sections). What perhaps they have been concerned
with is the boundary layer growth along the cyclone wall. The slower moving gas
closer to the wall shows diminishing Γ with increasing radius, and within a
certain narrow range of r (not too close because of the dampening influence of
the wall),the inner layers of faster-moving gas can spontaneously exchange with
the outer slower layers, doing so in "packets", creating turbulence. This then
will be the overriding mechanism for transferring angular momentum outwards towards
the wall, described by the "Reynold's stress".

$$\tau_{Tr\theta} = -\rho \overline{V_r' \, V_\theta'} \qquad\qquad\qquad (20)$$

where the primed velocities are the fluctuations about their averages, and the bar
denotes a time average. Instead of radial motion of individual molecules (by
diffusion) carrying ρV_θ momentum, as in the laminar case, we have now radial
motion of *aggregates* of molecules.

If turbulence has been introduced, the $V_\theta(r)$ profile is expected to be governed by the Reynolds stress, rather than viscous stress. The introduction of turbulence often occurs with the incoming fluid flow, for conditions in the inlet duct are often turbulent with $Re_{duct} > 2000$. This appears to be the case with a study of water flow by Ito, Ogawa and Kuroda (ref. 32). Their double tangential entry had a Re within the entry ducts of at least 6.5×10^4, and they observed **turbulence** dominated flow within the tube of rotating fluid, for Re_z as low as 5000, even though their V_θ had the Burgers vortex form of eqn. (14), and their Γ increased until close to the wall. Their Ω was an order of magnitude larger than that of Faber and Leibovich (ref. 29). A major difference between the two studies causing turbulence may also lie in the quality of the tube surfaces. That of Faber and Leibovich was machined and highly polished, (and thus closely round) whereas that of Ito, Ogawa and Kuroda was apparently that of as-extruded plastic pipe. Generation of turbulence close to the wall may influence a rather startling finding from their study. They found that the Reynold's stress $\overline{\rho v_r' V_\theta'}$ diminished to zero at the radius r_b where V_θ is a maximum. Fig. 5 shows some of their measurements. Their results deserve some study, because their conditions are close to those of many industrial cyclones, and because of the possibility that even in turbulent flow, the momentum transfer or drag between the inner core and the outer vortex may depend only on the viscosity.

The conservation of θ momentum with turbulent flow takes the form

$$\frac{\partial \overline{V}_\theta}{\partial t} + \overline{V}_r \frac{\partial \overline{V}_\theta}{\partial r} + \frac{\overline{V}_r \overline{V}_\theta}{r} = \nu \frac{\partial}{\partial r} (\frac{1}{r} \frac{\partial}{\partial r} (r\overline{V}_\theta)) + \frac{1}{r^2} \frac{\partial}{\partial r} (r^2 \overline{V_r' V_\theta'})$$

$$- \overline{\rho' V_r'} (\frac{\partial V_\theta}{\partial r} + \frac{V_\theta}{r}) \tag{21}$$

where the bars denote time averaged values, and we have retained the possibility of fluctuations in density ρ' for the turbulent motion (see Hall, ref. 32). The magnitude of the viscous and second terms on the right hand side for the conditions stated in Fig. 5, are -4×10^{-4} m s^{-2} and -3.5×10^{-3} m s^{-2} respectively (for $0.8 \, r_b < r < 1.2 \, r_b$) but the third term is unknown. It is expected to be considerably smaller than the normally considered $\overline{\rho V_r' V_\theta'}$ term. Ito, Ogawa and Kuroda (ref. 32) assume from their experiments that the viscous term of this equation is the controlling one, for r close to r_b. On this basis, they say that the situation at r_b is similar to that close to a wall, with $\overline{V_r' V_\theta'} \to 0$ as $r \to r_b$. They take the total shear stress $\tau_{r\theta}$ as the sum of the terms

$$\tau_{r\theta} = \mu (\frac{\partial \overline{V}_\theta}{\partial r} - \frac{\overline{V}_\theta}{r}) - \overline{\rho V_r' V_\theta'} \tag{22}$$

$\partial V_\theta / \partial r$ is also zero at $r = r_b$, so the shear stress is simply the second term, and

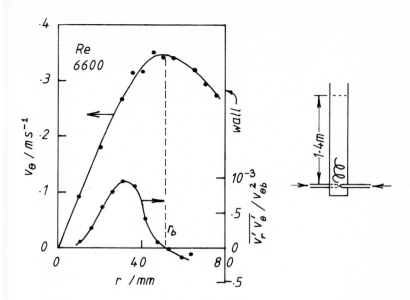

Fig. 5. Turbulent vortex in a tube

a characteristic "friction velocity" $V_{\theta f}$ may be defined

$$V_{\theta f} = \sqrt{\frac{\tau_{r\,\theta,\,b}}{\rho}} = \sqrt{\frac{\nu \overline{v}_{\theta b}}{r_b}} \qquad (23)$$

One can then make the whole problem dimensionless, by relating distances to r_b and velocities to $V_{\theta f}$. This is an attractive possible simplification and generalisation for cyclone flow. It is reasonable that the Reynolds stress should vanish at r_b if the turbulence is generated by instability between circulating shells of fluid with differing V_θ, and this condition is probably fulfilled in Ito, Ogawa and Kuroda's study, since their Ω number was high (7 to 20). However, some care should be taken, for larger scale turbulence not measured by Ito, Ogawa and Kuroda could still contribute to radial momentum transfer, and the third term in eqn. 21 involving $\overline{\rho'\,V_r'}$ may also contribute.

The integration of the turbulent eqn. 21 is not expected to yield the simple Burger's vortex profile 14, but the experimental profiles in Fig. 5 are suffici-ently close to this form to consider eqn. 14 as a useful expedient. Hall (ref. 31) has summarised attempts to fit the Burger's profile to turbulent swirling flow by assuming a constant eddy viscosity ν_T to apply across the whole flow. He shows that an effective ν_T (derived from a "best fit" to Burger's profile) may be correlated with moderate success by.

$$\frac{\nu_T}{\nu} = \Lambda \left(\frac{\Gamma_o}{\nu}\right)^{\frac{1}{2}} \qquad\qquad\qquad (24)$$

where Λ is a "constant" which varies from 0.2 to 0.6 for Re $= \Gamma_o/\nu$ from 2 x 10^3 to 10^7. This result was obtained from measurements on unconfined trailing vortex cores, where A can be estimated.

2.3 Vortex breakdown

A vortex develops a lower pressure at its centre, and it is one of the more striking demonstrations of boundary-free flow that a fluid will stopper the ends of its vortex motion. Leonardo da Vinci briefly summarised much of his early observations on vortices when he wrote "of the eddies one is slower at the centre than the sides, another is swifter at the centre than on the sides; others there are which turn back in the opposite direction to their first movement". He was (apparently) in the last phrase talking about the axial velocity.

As the strength of swirling motion increases, many swirling flows suffer a reversal of axial flow towards their centre, opposing the axial drift of the outer layers. Harvey (ref. 28) was the first to realise that the transition between like and reverse core flow was marked by the so-called "vortex breakdown phenomena". From his experiments with air in a tube, using a smoke tracer, he could best describe the effect as that of "an imaginary body of revolution placed on the axis of the vortex, around which the fluid is obliged to flow". He found that the imaginary object took the form of a hemispherical bubble of almost stationary fluid, headed by a free stagnation point. The position of this "bubble" was extremely sensitive to disturbance, and it could be triggered *upstream* of a disturbance - e.g. a probe inserted into the flow. Vortex breakdown occurs whether conditions upstream are laminar or turbulent, and downstream flow is always turbulent, with generally a larger scale (Hall, ref. 31), and often a spiralling further disturbance. Faler and Leibovich (ref. 33) have investigated the structure of such a breakdown, and found a two-celled internal flow, with altogether four stagnation points on the axis. Fig. 6 is a representation of their velocity measurements and also shows an instantaneous shape. The bubble has been found to fill and empty from the rear, and Fig.6 shows the diametrically opposed emptying and filling points observed by Faler and Leibovich (ref. 29), which in themselves were rotating.

The occurrance of a vortex breakdown has been found to depend on the magnitude of V_θ/V_z (or Ω) and only weakly on the Re_z value. It appears to be induced by configurations encouraging an adverse dp/dz - thus divergent ducts have a lower transition Ω than do parallel walled pipes. The emergence of a swirling jet into an unconfined space (as in the important example of swirl burners) however has a vortex breakdown at a little higher Ω. Thus Harvey found vortex breakdown for a

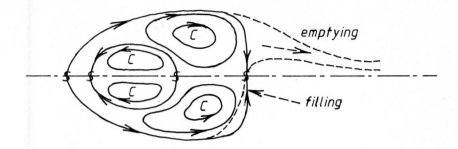

Fig. 6. Details of a vortex breakdown bubble. Solid lines-mean streamlines, dotted lines instantaneous bubble profite at rear. S stagnation points, C circulation centres.

helical angle $\phi \simeq \tan^{-1} V_{\theta}/\overline{V}_z$ of 20° at the wall $(\Omega = 1.1)$ in a regular pipe, Faler and Leibovich note a transition for $\Omega = 1.07$, but swirling exhaust from a tube into a large furnace cavity begins recirculation (reverse axial flow) for $\Omega = 1.9$. Two recent reviews (Syred and Beer, ref. 34, and Lilley, ref. 35) give an outline of the many measurements made on swirling flows for combustion, some of them pertinent to this discussion.

Beyond the formation of a stagnation bubble, for higher Ω the stagnation bubble moves upstream along the centreline, often followed by an asymmetric disturbance. A continuous reverse flow then sets in, converting what may have been a laminar core to a highly turbulent one. the spiral instability often seen downstream of the stagnation bubble (e.g. Harvey ref. 28, Faler and Leibovich ref. 29) now seems to have its counterpart in a reverse-flow "precessing vortex core" of importance to swirling combustors. The centre of core rotation lies off the axis of the outer flow, and follows a helical path close to the boundary of the mean reverse flow core. This helical path precesses about the axis at a frequency which is in the range 15-200 Hz and is roughly proportional to flow rate in a given device (Syred and Beer, ref. 34). The movement of this coherent fluid structure appears to dominate at least the larger scale turbulence in the reversed core.

A clear picture of the development of larger relatively unconfined swirling flows can be gained from Church, Snow, Baker and Agee's experiments (ref. 36) with air in the tornado simulator shown schematically in Fig. 7. They could generate an almost ideal potential vortex in this device and with radial inflow and upward axial flow achieve a laminar core. They discuss the phenomena observed in terms of three dimensionless groupings, obtained by making the Navier-Stokes equations

20

(momentum conservation, with viscosity) dimensionless. If the volume flow per
unit height in the convergence region is denoted as Q, so that the total flow is Qh,
then the ratio of angular to radial momentum is Γ/Q, and the three dimensionless
numbers appearing in the above equations are

$$
\begin{aligned}
S &= r_o\Gamma_o/2Qh \\
Re_r &= Q/2\pi\nu \\
a &= h/r_o
\end{aligned}
\tag{25}
$$

r_o and h are dimensions defined in Fig. 7, Re_r is a radial Reynolds number, and
S is a swirl number analogous to Ω.

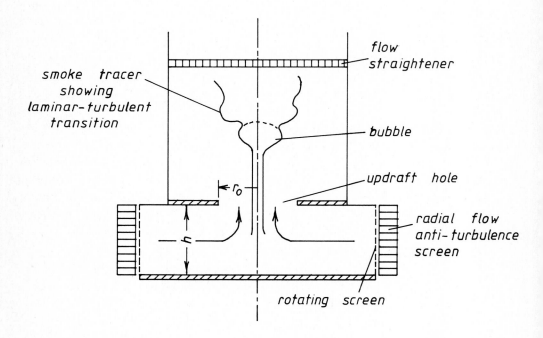

Fig. 7. Tornado simulator

A vortex breakdown occurred for $S \cong 0.3$, and as the Reynolds number was large
($Re_r > 5 \times 10^4$) the laminar core became permanently turbulent, also with a greatly
enlarged diameter. The maximum upward velocity was in an annular ring about the
breakdown bubble. As the swirl was increased, the breakdown moved against the
stream with the turbulent core following, until at $S \sim 0.5$ the breakdown bubble

neared the lower surface. Sometimes downflow did not follow closely on the
breakdown bubble, in which case a single spiralling roll vortex (S > 0.4)
(stationary about the vertical axis) was clearly seen in the shear layer between
the core and surrounding upflow. When the central downflow penetrated to the lower
surface (S > 0.5) it was surrounded by an upward moving core, within which were now
a pair of intertwined helical vortices (now rotating themselves about the central
axis). With expansion of this upward moving core as S was increased, as many as
six subsidiary vortices (still apparently regularly arranged) were observed by
photography and tracer methods. The transitions described above depended mainly
on the value of S, and secondly on Re_r, lower asymptotic transition S values being
reached for higher Re_r (3×10^5), and being almost independent of the aspect ratio
a.

When the radius of the core was measured (defined by the position of maximum
mean total velocity, measured by hot film techniques) it was found to be proportional to S (up to S = 2, when the influence of the size of the updraft hole
began to be important) and independent of Re_r and a (Baker and Church, ref. 37).

2.4 Application to Cyclone Flows

In applying the observations made on "unconfined" rotating flows to cyclones, care
must be taken to distinguish those features which are characteristic to the basic
sink swirling flow, and those which are specific to the detailed wall geometry.
In one sense, we can envisage dust-collecting cyclones as devices which produce a
sink vortex, and which then interfere more or less with the properties of the
vortex, depending on the detailed design. We also may need to discuss separately
small and large cyclones. We can begin by considering cyclones of intermediate
size (0.1-1 m dia.), and those designs for which high dust collection efficiencies
have been found.

2.4.1 Tangential velocity

In tangential return-flow cyclones which have been "well designed", the maximum
tangential velocity V_θ appears to be maintained or even increased as one travels
along the cyclone towards the dust exit. Thus ter Linden (ref. 18) found a maximum
of 20 m s^{-1} increasing to 30 m s^{-1} towards the bottom of a long cone cyclone,
compared with the inlet velocity of 10.7 m s^{-1}. His "circulation" $\Gamma = V_\theta r$ dropped
from the inlet to the maximum V_θ position by about 60 percent, indicating non
potential flow. The shape of the V_θ versus r profile is similar to that of
Burger's vortex. In one set of measurements on a cylindrical cyclone (D = 140 mm,
without a cone), Smith (ref. 38) has given detailed plots of Γ versus r for
different axial locations. Away from the gas exit, this author has found that his
profiles could be matched well by Burger's vortex

$$\Gamma/\Gamma_R = 1 - \exp\left[-C(r/R)^2\right] \tag{26}$$

with the constant C = 11 (see Fig. 8). (R is the cyclone wall radius and Γ_R is the circulation at the wall). Γ_R dropped only 15 percent along his cyclone, of length to D ratio 2.5. It is interesting also that ter Linden's velocity profiles can be described by eqn. 26 (at least from his wall and peak velocities) with C = 11.6, ~14, ~6 for successive stations down into the cone of his cyclone. Γ_R did not vary more than the accuracy of reading (10 per cent) down the wall, except close to the dust exit. Also some profiles in a long cone cyclone (D = 300 mm) published by Pervov (ref. 39) could be described with C = 9, 13 (these are within reading accuracy of each other). Hejma (ref. 40) has given some more easily read V_θ versus r curves for a D = 314 mm cyclone, and fitting wall and peak velocities within the cylindrical section, C = 9.6. The value of C appears to be a common value of about 10 for cyclones of differing geometries (D_e/D for the cyclones mentioned above varied from 0.35 to 0.54, L/D = 2.2 to 6.0 where D_e is the gas exit diameter and L is the total axial length from the roof to the dust exit), and may have some use in particle collection calculations. A more careful check with a wider range of cyclones is recommended. We can note here that in the normally used correlation for V_θ for $r > r_b$,

$$V_\theta/V_{\theta R} = (R/r)^m \tag{27}$$

Alexander (ref. 41) has found by measurements of cyclones with D = 25 mm to 1.2 m that m increases weakly with D (to the 0.14 power). Also he found that m diminished with temperature (1-m was proportional to the absolute temperature raised to the 0.3 power). Kinematic viscosity is inversely proportional to the 0.5 power of absolute temperature T at constant pressure; thus from both the D and T dependence, it appears that m has a weak increasing dependence on a Reynolds number. Since our relationship eqn. 26 can be expressed

$$V_\theta/V_{\theta R} = \frac{R}{r}\left(1 - \exp\left[-C(r/R)^2\right]\right) \tag{28}$$

i.e. also as a function of r/R, then it appears that C must be a function of Reynolds number, and we can suggest that C is a *weakly* increasing function of Γ_R/ν.

For most purposes however, C ≃ 10 may be a useful empirical value for determining the size of r_b. All previous workers have related this core size to the gas exit diameter. What is being suggested here is that r_b depends on the basic swirl and sink properties of the cyclone vortex away from the gas exit, and depends on the barrel radius rather than the exit radius. In fact r_b/R can be calculated by taking $dV_\theta/dr = 0$ in the Burger's vortex, resulting in

$$C\left(\frac{r_b}{R}\right)^2 = 1.25 \qquad\qquad (29)$$

so that for $C \simeq 10$, $r_b/R = 0.35$

2.4.2 Radial velocity

The radial velocity, although small and difficult to measure, is of paramount importance in the collection or loss of small particles, as will be discussed in section 3. A major difficulty of allowing in design for the effect of radial velocity has been that measurements made have been few and at limited locations in a cyclone; cylindrical symmetry and axial uniformity is not guaranteed especially close to the gas entry and exit. Ter Linden (ref. 18) reported V_r to be $1-2$ m s^{-1} (in flow)(cf. $V_i = 10.7$ m s^{-1}) over most of the length of his cyclone and for the outer radii. For small radii V_r was outward, and the transition between inflow and outflow was somewhat greater than r_b. Hejma (ref. 40) measured $V_r = 1 - 3$ m s^{-1} for $V_i = 17$ m s^{-1}, whereas Pervov (ref. 39) measured $1 - 2$ m s^{-1} reaching about 3 m s^{-1} 0.5 D_e below the gas exit. Razgaitis, Guenther and Bigler (ref. 42) taking what appear to be more accurate measurements in a larger cyclone ($D = 914$ mm) find at a single axial position 1.6 D_e below the gas exit, an inward $V_r = 2.0 - 2.5$ m s^{-1} from 0.2R to 0.7R, and an outward $V_r > 1.5$ m s^{-1} for $0.75 - 0.95$ R (cf. inlet velocity $V_i = 13$ m s^{-1}). These "spot" measurements can be compared with the average radial inward velocity calculated by Barth, averaged over his cylindrical "collecting surface" (1.1). This average V_r is 0.35 m s^{-1} for ter Linden's cyclone and 0.45 m s^{-1} for that of Hejma. It is on these much lower velocities that Barth's design is based.

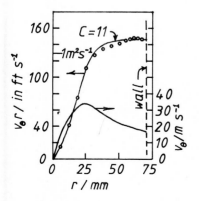

Fig. 8. Fitting of Burger's vortex $\Gamma/\Gamma_R = 1 - \exp\left[-C(r/R)^2\right]$ to profiles in a return flow cyclone (also showing V_θ).

We can perhaps improve on Barth's approach (which assumed that $V_r \propto V_i$) by allowing for some scaling of the sink flow according to the Burger's vortex model. (Note that any scaling other than $V_r \propto V_i$ will have to take a non-uniform V_r over the height of the cyclone).

The coefficient C can be expressed as

$$C = AR^2/2\nu_T \qquad (30)$$

where A describes the level of inward radial sink flow in the cyclone, and ν_T is expected to increase with vortex Reynolds number Γ_R/ν, perhaps as in eqn. 24 for an unconfined vortex. Γ_R/ν varies from 3×10^4 to 2×10^5 in the above studies, with a 2.5 fold increase in ν_T expected from eqn. 24. For the value of C to remain effectively constant, the level of radial sink flow must also increase with Γ_R/ν. C can be seen as a **radial** Reynolds number, with AR a *characteristic* radial velocity. Evaluating ν_T from eqn 24 (for an alternative, see 3.7), the characteristic radial sink velocity becomes

$$V_R = AR = \frac{2C\Lambda(\Gamma_R\nu)^{\frac{1}{2}}}{R} = 2^{3/2}C\pi^{\frac{1}{2}}\Lambda\left(\frac{V_{\theta R}\nu}{R}\right)^{\frac{1}{2}} \qquad (31)$$

where $V_{\theta R}$ is the tangential velocity extrapolated to the wall, ignoring the boundary layer. In evaluating V_R we might use the dimensionless constant $C \simeq 10$ from above, and $\Lambda = 0.4$ from measurements on trailing vortices. A value of Λ more appropriate to cyclones may be calculated from measurements of the Reynolds stress $\overline{V_r' V_\theta'}$ (as Razgaitis, Guenther and Bigler have done) together with a V_θ - r profile Thus we could use

$$\nu_T \simeq \frac{\overline{V_r' V_\theta'}}{\partial V_\theta/\partial r} \qquad (32)$$

to evaluate ν_T. Unfortunately their $\partial V_\theta/\partial r$ values are not available. Values of ν_T will be estimated later in 3.7 from a *confined* vortex model. The absolute value is not important, but the scaling of V_r according to $(V_{\theta R}\nu/R)^{\frac{1}{2}}$ or $(V_i\nu/R)^{\frac{1}{2}}$ provides an extreme (unconfined vortex) alternative to previous scaling according to V_i.

The above approach assumes that we have an ideal cyclone design - i.e. the radial sink flow is no more than that required by a free-standing sink vortex with the turbulent viscosity ν_T found in such a vortex. However the bounding walls require us to satisfy the total volumetric flow by radial flow along the cyclone length, so actual velocities will be different from those characterised above, for several reasons. Abrahamson, Martin and Wong (ref. 43) have shown by velocity measurements in the bin of a Stairmand design cyclone that about 10 per cent of the total gas

flow circulated into the bin before rising centrally into the core. Pervov
(ref. 39) has also found that 10-15 per cent of the gas enters the dust collector.
This gas bypasses the normally considered radial flow; more may do this within
the core, close to the dust exit.

The most revealing measurements for V_r are those derived from **differences
between total downward gas flows, measured (by V_z) at different levels of** z.
Reznik and Matsnev (ref. 44) have done this for three return-flow cyclones
of D = 230-254 mm, and found that the total vertical gas flow diminished sharply
just below the gas exit section. In only 25 mm from the level of the gas exit,
the decrease was 42, 51 and 32 per cent for cyclones with axial vaned entry,
scrolled (tangential) entry and semi scrolled entry respectively. These drops
corresponded to average V_r within the 25 mm of 6.7, 7.6 and 4.5 m s^{-1} respectively.
Further down the cyclones V_r was much lower (0.4-0.8 m s^{-1}). These startling
values are not peculiar to their cyclones, for Smith (ref. 38) finds by the same
technique a drop of 53 per cent of axial flow in the first 25 mm below the gas
exit. He notes a major effect of the height of the inlet opening (his entry is
tangential, through 8 vanes). The above result was found for an inlet height of
0.36D, whereas for a smaller height of 0.089D essentially no radial inflow was
found over the first 50 mm, and thereafter the radial velocity was close to 0.3
m s^{-1} over most of the length, as found with the cyclone with larger height.

It appears that the "end effects" of gas entry, gas exit and dust exit can be
of major importance when considering effective radial velocities, and the direct
measurement of radial velocity may give unreliable or misleading information.

2.4.3 Axial Velocities and Vortex Breakdown

Smith (ref. 45) observed with **smoke** visualisation studies in his cyclone that
when the flow of gas was suddenly reduced, **an axially** symmetric flow-adjustment
wave travelled down from the gas exit and reflected off the flat bottom. Alternately, an asymmetric disturbance caused a helical disturbance to travel down.
Ayer and Hochstrasser (ref. 23) have reported two distinct flow regimes for their
small (D = 19 mm) cyclone, if the cone was long. One of these regimes resulted in
the deposition of dust in a regular ring part way up the cone, and in no further
change in measured pressure below this "transition" level. This regime they termed
"laminar" as it was associated with an Re_z in the exit tube of <2000. At higher
flows they observed normal dust collection behaviour, and they called this the
"turbulent" mode as it occurred for an Re_z > 4000 in the exit tube. For 2000 <
Re_z < 4000, either mode could occur. Only the turbulent mode was observed for
short coned cyclones.

It appears that Ayer and Hochstrasser observed a vortex breakdown near the
bottom of their cyclone, and correctly associated it with upstream disturbances. The

value of Ω in cyclones defined formally as in eqn. 18 is

$$\Omega = \frac{2\pi R_i V_i}{D \overline{V_z}} = \frac{2\pi R_i Q}{D A_i Q} \frac{\pi D^2}{4} = \frac{\pi^2 R_i D}{2A_i} \tag{33}$$

where Q is the total volumetric flow rate, A_i is the inlet cross sectional area
and R_i is the radius of the mid point of the inlet area. Thus it is simply a
ratio of areas, and has the value 19.7 for the Stairmand cyclone design that Ayer
and Hochstrasser used. Effectively we expect that if their cyclone was continued
in its cylindrical section, (Ω in eqn. 18 was defined for swirling flow in a
tube) a vortex breakdown and strong reverse flow would have occurred since $\Omega \gg 1$.
A conical section and converging flow will guard against an adverse pressure
gradient in the axial direction, which could precipitate a vortex breakdown (ter
Linden (ref. 18) shows a diminishing static pressure, and Pervov (ref. 39) shows
a constant static pressure with distance into the cone). However the abrupt
enlargement of flow area into the bin will precipitate breakdown and locate the
position of the reverse flow. At such a high Ω value the reverse flow is expected
to be strong,causing recirculation in the bin, and the primary recirculation bubble
will migrate to the gas exit. It is expected that the cyclone will normally
operate in a regime beyond the laminar vortex core well into a single or multi
spiralling vortex core system, as shown clearly by Church, Snow, Baker and Agee
(ref. 36) in their Fig. 8. However, they also show there that for **lowered Reynolds**
numbers, the critical swirl number (analogous to Ω) for transition from the laminar
core rises markedly. It seems plausible that Ayer and Hochstrasser operated at
a sufficiently low Reynolds number to avoid this transition, and the vortex
breakdown appeared in its laminar form. The explanation offered here for normal
"turbulent" operation of dust collecting cyclones would seem quite general, for
most cyclones will have $\Omega > 10$.

It is now of interest to examine the gas exit closely. Ter Linden shows an
axial upward velocity in the core flow entering the gas exit, which is reduced in
the centre. Smith (ref. 38) also shows a reduction near the centre and for his
measurements closest to the gas exit shows a *downward* axial velocity on the axis.
These velocity profiles cannot be the result of drag interaction with the wall,
and can be easily explained by the existence of a vortex breakdown bubble close to
the lip of the exit pipe. Pervov (ref. 39) in fact reports a stagnant zone in the
shape of a cone in the centre of the cross-section of his exhaust pipe, with the
base of the cone turned towards the flow. The gas leaving the cyclone passed
through an annular space between the tube wall and the stagnant core, with high
axial velocities.

Some more insight into the "turbulence" in the core observed by several workers
with tracers (e.g. Stairmand ref. 1) and hot wire anemometers (e.g. Schowalter and

Johnstone ref. 46) may be gained from Smith's high speed photographs of the
movement of smoke tracer in the core. He found that a single vortex existed in
his cyclone (Ω = 7.4); a vortex which had a rapidly moving axis of rotation, and
gave the impression of general turbulence to the eye. At the higher velocities
the vortex had a clear core (probably from the clean boundary layer flow induced
at the base of his cyclone, out of the smoke injection region) which remained a
clear core over the length of the cyclone. This indicates very little fine scale turbulence within the vortex, and that the vortex retains the gas flowing up through it
until reaching the exit.

Smith's generation of spiral   disturbances by partially blocking off the gas
exit from the side (see above) suggest how the instabilities in position of the
vortex might arise. Vortex breakdown bubbles are notoriously unstable in radial
position, and easily wander away from the centre, effectively blocking off one side
of the exit more than the other. The eccentricity in position of the breakdown
bubble will then cause the vortex to wander from its position on the axis of the
cyclone. In support of this, Smith found that small eccentricities of only 0.4 mm
in the exit tube were enough to reverse the departure from symmetry of the measured
axial velocity profiles. Thus only minor motions of the breakdown bubble will
cause large asymmetric movements of the vortex.

There appears to be another cause for periodic motion of a single cyclone vortex,
affecting the lower part of the core. If the cyclone is long (L/D > 4) the lower
vortex end, normally operating in the dust exit, can attach itself to the cyclone
wall further up, and move around at frequencies found by Smith (ref. 38) to be a few
Hz at low flows to 40 Hz at higher flows. Smith found that this whipping vortex
end completely dominated the flow in the lower cyclone, and Alexander (ref. 41) has
found in this instance a stagnant zone between the vortex end and the dust outlet.
In terms of our vortex theories discussed so far, this phenomenon can be described
as a preference for vortex breakdown and reverse flow at the wall rather than in the
dust exit-bin geometry.

A simple argument may help to understand this transition, which is important for
cyclone design. If we ignore the loss of angular momentum (or circulation Γ_R)
as the gas moves down the wall, for a cyclone without a cone the loss of gas through
radial drift will result in a reduction in axial velocity until it is zero, at
which point there is no axially symmetric feed for the reverse flow in the centre.
Without this stabilising feed, the vortex will seek the easiest supply of gas for
its low pressure centre, taking it from the wall boundary layer, with also a V_θ
component of momentum, resulting in a rotation of the vortex end around the wall.
If the cyclone has a cone, the axial velocity is maintained for a longer axial distance, because of the smaller cross section.

The maximum "natural length" of a cyclone vortex before attachment to the wall

has been reported by Alexander as being independent of "velocity" over a wide
range, inversely proportional to the cube root of A_i/D^2, or Ω , and longer for
larger exit tubes, presumably where the effect of small radial movements of the
gas bubble is not so important in causing asymmetric flow further down. These
asymmetric movements of the vortex are likely to contribute strongly to radial
transfer of momentum and hence in turn affect v_T. With a smaller v_T (for larger
exhausts?) the radial sink flow will decrease (e.g. by eqn. 27) and hence the
"natural length" **may** increase.

2.4.4   Pressure drop

The total pressure drop ΔP across a cyclone is often the major factor affecting
the cost of the dust collection, and many workers have attempted to correlate ΔP
with the geometry. By ΔP we mean the difference in head $(P_{stat} + \rho v^2/2)$ between
the inlet and outlet. Löffler and Meissner (ref. 47) have argued that the
location of the outlet measurement cross section should be carefully chosen to yield
a full measure of ΔP. Because of the spin in the exhaust gas, they recommended a
measurement location some 150 D_e downstream from the exit (which is clearly not
possible for many cyclones). Thus,as many measurements must be done close to the
cyclone body, the cyclone ΔP we find in many publications may be lower than this
ideal. This again may not disturb us, because the conversion of the spin kinetic
head to axial kinetic head or to static pressure P_{stat} (as required for pumping
along a pipe) is not likely to be efficient, and we may be more interested in
$(P_{stat} + \rho v_z^2/2.)$

Two major efforts in describing contributions to ΔP are those of Stairmand
(ref. 48) and Barth (ref. 4) as discussed by Strauss (ref. 49). Both descriptions
are empirical to a large extent, but both break the total ΔP into various components,
of which that associated with the entry and exit are dominant. Other components
are frictional losses in the entry duct, near the cyclone walls and in the exit
pipe.

One interesting treatment of pressure drops is that of Alexander (ref. 41) who
considered that change in $(P_{stat} + \rho v_\theta^2/2)$ suffered by gas in moving from the
outer layers of a $V_\theta \propto r^{-m}$ vortex to the inner layers, and hence to the exit. The
rotational head lost with the exit gas was calculated considering a solid rotating
core **with** diameter $D_e/2$. He found that the exit loss (amounting to the rotational
head, considered not recoverable) was over 80 per cent of the total for a typical
150 mm cyclone, and that close correspondence with experiment was obtained.

In considering a different class of cyclone, Tager (ref. 50) has had success in
correlating the pressure drops associated with a variety of geometries used for
combustion chambers. He divides the pressure drop into that associated with the
entry and with the cyclone **chamber** proper. Of this latter, he found evidence that

the exit loss dominated. His flows were sometimes different at inlet and outlet and so he perhaps made a clearer distinction between inlet and cyclone chamber-exit contributions. His entry contributions ΔP_i of relevence here are in the form

$$\Delta P_i = \zeta_i \frac{\rho_i \, V_i^2}{2} \tag{34}$$

where ζ_i was 1.3-1.4 for well profiled scroll (tangential) entries, and 1.2 for axial entry through swirl vanes. Tager's cyclone chamber pressure drops ΔP_{cc} were expressed as

$$\Delta P_{cc} = \zeta_{cc} \frac{\rho_e \, V_{ez}^2}{2} \tag{35}$$

where ρ_e is the exit gas density and V_{ez} is the average axial velocity of gas in the exit. ζ_{cc} diminished with the exit axial Reynolds number until it remained constant at 4.0 for a reverse flow cyclone ($Re_{ez} > 10^5$) with $D_e/D = 0.45$. In order to estimate ΔP_{cc} for other size ratios, the expression

$$\frac{\Delta P_{cc}}{\Delta P_{cc \; 0.45}} = \frac{0.9D}{D_e} - 1 \tag{36}$$

was found to correlate return and through-flow geometries successfully(operated with hot and cold exit flows)within $0.3 < D_e/D < 0.7$.

Tager's expression applies to short cyclones ($L/D \approx 1$) in which several complete V_z flow reversals occur across the diameter (see Syred and Beer ref. 34, and Baluev and Troyankin ref 51). His ΔP_i is close to   Stairmand's recommendation for the inlet contribution (one inlet velocity head loss) but his exit contribution is much too low when applied to dust collection cyclones (e.g. contributing only an extra 1.3 and 1.7 to ζ for cyclones A, B in Fig. 1). His results appear to be reliable, for Syred and Beer have found them to predict an accurate ΔP for burners they have measured (ref. 34). Tager has clearly illustrated the effect of changing D_e/D, but his measurements were taken on cyclones with perhaps two characteristic differences from most dust collection cyclones. The lengths were short, and the characteristic swirls were many times less.

We can understand some of the gas movement which may take place just inside the gas exit of a dust collecting cyclone, when we study the observations made on swirling burners at lower swirl. Syred and Beer (ref. 34) have shown that for Ω in the range 1-5, a higher Ω results in a larger diameter recirculation volume, with greater recirculation (up to 1.5 times the swirling mass flux is entrained from surrounding gas into the reverse central flow). Large increases in pressure drop with swirl strength were reported. Mathur and Maccullum (ref. 53) show for swirling air flows issuing from a burner nozzle 98 mm dia. into a square duct of size 244 mm,

that for *nozzle* Ω values of 6 and above, the central recirculation flow had
become large enough to limit the forward flow to the outer 50 per cent of the
cross section. It is to be expected that in a continuing circular duct as in a
cyclone exhaust, the restriction at the same Ω would be more severe. Syred and
Beer show that pressure drops for swirl burners are about 50 per cent of the swirl
kinetic energy (for Ω = 2-3). The pressure drop across the breakdown bubble in
a cyclone exhaust (at larger Ω) is expected to be at least this magnitude.

From these descriptions, and in view of Alexander's success at relating pressure
drops to swirl energies, one might expect that the major pressure drop in a
cyclone will be shown across the breakdown bubble in the exhaust, between the
forward stagnation point and perhaps the rear stagnation point. Pervov (ref. 39)
in fact shows that a static ΔP equal to the pressure drop across his cyclone,
occurred within the gas exhaust, on the centreline, between stations 0.7 D_e and
3.4 D_e into the exhaust. This finding illustrates vividly the role the gas bubble
plays as a stopper for the vortex. However as well as being physically necessary
to maintain the central vortex vacuum, the bubble is expected to influence
the magnitude of ΔP in two other ways. The stability of the bubble-exhaust tube
annular flow (dependent on swirl, lip geometry, regularity and roughness) is likely
to influence the movement of the vortex core, and hence radial "diffusion" of
angular momentum. The balance of diffusion and convective flow of this momentum
is described by the radial Reynolds number C, which appears to have a similar
value for different cyclones. The larger each of these flows, the lower the swirl
is expected to be at the exit, and the lower ΔP is expected to be. The other
influence of the bubble flow on the pressure drop is via the bypassing of low-
swirl gas up the annulus, as mentioned above in 2.4.2. This route for the gas
constitutes essentially a short circuit, and the overall ΔP will fall.

In correlating ΔP we might then look for (a) a measure of exit swirl, and (b) at
the geometry of inlet (which appears to determine the extent of gas short-
circuiting). Gas short-circuiting probably occurs to a limited extent in most
cyclones, depending on low-angular-momentum gas moving down radial pressure
gradients across the roof, and down the outside of the exhaust. Bloor and
Ingham (ref. 53) have applied boundary layer theory to the flow just beneath the
roof of a cyclone, and estimated that of the order of 10 per cent of the total gas
flow will leak inwards.

We can define a swirl number Ω at the exit

$$\Omega = \frac{\pi V_{\theta e}}{V_{ze}} \tag{37}$$

where $V_{\theta e}$ is the swirl velocity at the exhaust lip, and V_{ze} is the average axial

exhaust velocity. If we take $V_{\theta e}$ to be the swirl velocity from the Burger's
vortex eqn. 28, and calculate V_{ze} from a flow balance through the cyclone $V_i A_i =$
$\overline{V}_{ze} A_e$, then

$$\Omega_e = \frac{\pi A_e}{A_i} \frac{D_i}{D_e} (1 - \exp \left[-C(\frac{D_e}{D_i})^2 \right]) \qquad (38)$$

where $D_i = 2R_i$*. The pressure drop will be related here to a kinetic head in the
exhaust rather than the inlet, and for convenience only, we will relate it to the
average axial **kinetic head.** If ΔP is the total cyclone pressure drop, we
define

$$\zeta_e = \Delta P / \frac{1}{2}\rho \overline{V}_{ze}^2 \qquad (39)$$

and plot values versus Ω_e for cyclones of widely differing geometries in Fig. 9.
It is seen that ζ_e increases with Ω_e, with most cyclone points lying within a
band, and the cyclone burners of Tager (point H) fall within the general scatter.

The scatter in Fig. 9 may be explained at least in part by variations in actual
Ω_e from that calculated. The major error here is expected to be that due to
slowing of swirl by gas short-circuiting. Where the short-circuiting is known, it
has been indicated on Fig. 9 as a percentage of total flow. The effect of say 50
per cent short circuiting of non-rotating flow (by conservation of angular momentum)
is to reduce Ω_e by 50 per cent. (This has not been corrected for in points I, K, J).
In support of this effect, Shepherd and Lapple (ref. 52) noticed that dust movement
at the inner exhaust surface was at a helical angle of 45° ($\Omega=3$) compared to an
angle "considerable less" in the inner spiral below the exhaust (greater Ω). Gas
short-circuiting may occur due to the axially symmetric   mechanism discussed
above, or due to a non-symmetric leak initiated by the (normally) single entry
to the cyclone. If most of the leakage under the exhaust lip is localised, spot
measurements of v_r at a specific θ may be widely misleading, and even calculations
of v_r based on integration of v_z may be in error, for v_z is normally measured in
one θ plane only, and it is assumed that v_z is cylindrically symmetrical. It is
expected that as the gas bubble moves from its axis, or with **non**circularity of
the exhaust tube, localised leakage (whose position may rotate) is likely.

Gas short-circuiting offers reasons for a change of ζ_e with a change in the
inlet geometry shown in Fig. 9. It appears that cyclones with the inlet somewhat

---

* It has been found necessary for accurate calculation of V_θ to use the inlet
velocity V_i together with the radius R_i of the midpoint of the tangential gas
entry; for ter Linden's cyclone $L_i > D$, and $V_{\theta e}=20.3$ m s^{-1} calc. with C = 10 (cf.
22 m s^{-1} measured); for Hejma's cyclone $D_i < D$ and 24 m s^{-1} calculated cf. 26
m s^{-1} measured.

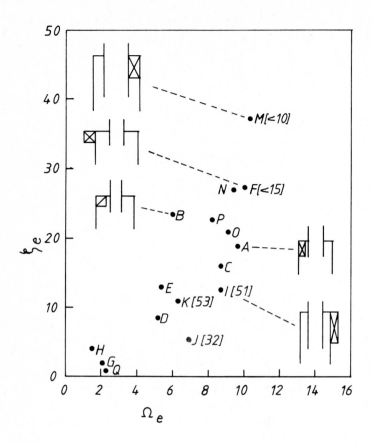

Fig. 9. Cyclone pressure drop number (based on exit velocity) versus exhaust
wall swirl number.

A. Stairmand Fig. 1.A.
B. Tengbergen Fig. 1.B
C. Stairmand Fig. 1.C
D. Tengbergen Fig. 1.D
E. Tengbergen ref. 2
F. ter Linden ref. 18
G. Shepherd ref. 52
H. Tager ref. 50 return flow
I. Reznick (b) ref. 44
J. Reznick (c) ref. 44
K. Smith I ref. 38
L. Smith II ref. 38 $\Omega_e = 16.8, \zeta_e = 85$, leakage=0
M. Hejma ref. 40
N,O,P,Q. Muschelknautz ref. 54
(measured core velocities used for Ω_e in K, L, M).

removed both radially and axially from the exhaust tube have higher ζ_e for a given
Ω_e than those where the inlet flow is close to the exhaust. Point I with 51 per cent
radial inflow in the first 25 mm below the exhaust lip can be compared with
point M, with less than 10 per cent radial inflow in the same region, and a more
shallow entry. The evidence in Fig.9 is still scanty, and this question should

be investigated more thoroughly, but this preliminary survey suggests that many designs may have considerable inflow just below the exhaust lip. It seems that the most sensible way to avoid this is to have an entry somewhat removed from the exhaust lip, and probably to have a closely circular exhaust cross section.

2.4.5 Cyclone wall instabilities

Coles (ref. 54) has investigated thoroughly the instabilities set up in a gas between two cylinders, the **inner** one of which is **rotating**   He shows many tanential periodic instabilities, and gives surface Reynolds number criteria for their occurrence. These periodicities began with a boundary velocity Reynolds number (of the inner cylinder, with radius 12 per cent less than the stationary outer cylinder) of about 1000 and finally distorted into irregular turbulence at a Reynolds number of 14000. The reason for these instabilities appears to be a gyroscopic precession of local rotation from an axial axis towards a tangential one, and it is likely that this will also occur in a cyclone along its inside walls. The above limits to periodic flow, when applied to a cyclone of 0.3 m dia. operating in air, correspond to tangential velocities between 0.07 and 1 m s^{-1}, so that instabilities would survive only within the boundary layer, where they should be observed in the patterns of dust flow.

Tani (ref. 78) describes steady periodic variations in tangential velocity measured close to a wall, curved concave in the direction of gas flow. These variations (periodic in the spanwise direction) occurred with both laminar and turbulent flow, and were a maximum within the boundary layer. Under Tani's conditions (air velocity 20 m s^{-1}, radius of curvature 5 m) velocity variations were of the order of 0.5 m s^{-1} at 5 mm from the wall in turbulent flow. The peak velocities were spaced about 70 mm apart across the flow. He related the variation to a system of alternating vortices with their axes in the direction of gas flow, proposed by Görtler. The effect of these Görtler vortices were predicted (and observed) to strengthen with smaller radius of curvature. In fact the Görtler number (G_T = 43 $(\theta/R)^{\frac{1}{2}}$ in Tani's paper, (θ is a **boundary layer thickness**), **which describes** the strength of the instability in turbulent flow, is an order of magnitude larger in most cyclones than in Tani's experiments.

---

3. DUST BEHAVIOUR

The collection of dust in a cyclone depends on its condition when it enters the cyclone. For this reason, we will take the reader through the dust history from the initial dispersion into air, and will deal with the details of dust movement until it is lodged on the dust heap in the cyclone bin, or is connected above the cyclone vortex breakdown.

3.1. Dispersal into the carrier gas

Most cyclone experiments are fed with dust from a hopper via perhaps a vibratory feeder,dropping into an air ejector whose vigorous air flow more or less disperses the dust. On the other hand, many industrial dusts are dispersed during their manufacture e.g. in grinding, or combustion or spray drying. They have often been selected from larger particles by elutriation against gravity.

First we will deal with dispersal in experimental studies. Kousaka, Okugama, Shimizu and Yoshida (ref. 56) have studied the break-up of particle agglomerates in air flows, where the size of their primary particles was at the lower limit we need to consider for cyclone collection, and therefore   the most difficult to disperse. They used $CaCO_3$ and Fe_2O_3 particles with geometric mean diameters 0.64 and 0.31 µm respectively, and studied agglomerates by direct observation through a microscope while samples were settling under gravity. Most of their work was done with a distribution of agglomerates formed by agitation at high dust concentration, with agglomerate diameters up to about 10 µm. They introduced a dilute flow of these agglomerates to various dispersing devices. Wire screens were shown to be most effective at breaking up the agglomerates at any given air velocity, showing the importance of impact forces, but at 8 m s^{-1}, the highest velocity used, most particles still remained in double or larger agglomerates. Of the other devices - venturi throat, orifice, capillary and rotating blade - poorer dispersion was obtained, with velocities around 30 m s^{-1} required to reduce the agglomerate size to 2 µm (perhaps 10 primary particles), and 100 m s^{-1} required for 1 µm. Their observations here reflect the difficulty of obtaining a high relative velocity of air and agglomerate, when the agglomerates are small to begin with. An improved dispersion was obtained when the agglomerated aerosol was fed laterally into the venturi throat, in a similar fashion to many ejector feeders, but even here, 120 m s^{-1} was required to achieve the same dispersion as with the screens. (See Yoshida, Kousaka and Okuyama ref. 57). From their experiments, and from theory, they conclude that impact on a solid obstacle is much more effective for dispersal than gas-particle forces and that for their conditions the latter are important in the order (a) acceleration in uniform flow (b) bending in simple shear flow (c) disruption in fine scale (Kolmogoroff) turbulence.

For the dispersal of larger particles, the study of Jimbo and Fujita (ref. 58) shows significant agglomerate break-up at lower velocities. They fed a variety of materials of primary size up to 55 µm, transversely into an air flow, along a 50 mm dia. tube, and collected the agglomerates on sticky tape for microscopic examination. Their agglomerates from a vibratory feeder were of the order 500-1000 µm in size, and were reduced markedly in size with increase in gas velocity, reaching 50-100 µm for some dusts at 30 m s^{-1}. However, extrapolation of their results to primary particle size indicated that air velocities of 100-300 m s^{-1} were required to achieve complete breakdown (to particles 7-15 μm size). They also showed some breakdown as the dust travelled along the tube. Agglomerate sizes reached steady state values after 1.5 m travel.

It can be inferred from the above experiments that it is difficult to break down fine dust into the primary particles by normally encountered duct flow. Even where **sonic** velocities (330 m.s^{-1}) are used in a jet ejector, and complete breakdown is probable, care should be taken with the dust so that it does not reagglomerate again in the turbulent flow as it is fed into the cyclone entry duct (see below).

Where dust has been collected by suction hoods from industrial sources, and has been carried along exhaust ducts for a considerable distance, it is pertinent to ask what agglomeration of the primary dust has occurred. (Most size analysis methods will more or less effectively disperse the agglomerates before analysis, and so this information, important to understanding collection mechanisms, will in many cases have been lost during sampling).

3.2 Agglomeration in a turbulent pipe flow

Fine particles can conceivably agglomerate in gas flow in a duct in two ways; either by colliding with each other directly and sticking, or by depositing on the walls and onto previously deposited particles, which are reentrained as agglomerates. We will consider direct collisions first, and then deposition and entrainment briefly.

A suspension of particles of less than 1 μm in size will suffer random Brownian motion, which leads to collisions between them. However in a turbulent gas, relative motion between the particles is also induced by the fine scale of gas motion, leading to collisions, and this turbulent collision rate is expected to dominate if *one* of the collision partners is greater than about 1 μm. For particles which are small and light enough to follow the finest turbulent motions, collisions are expected to occur between particles as they follow close streamlines past each other, *within the same viscous eddy* (see Fig. 10(a)). From Kolmogoroff's dimensional analysis of viscous dissipation of turbulent energy, the scale

$$\lambda = (\nu^3/\varepsilon)^{\frac{1}{4}} \tag{40}$$

roughly defines the size λ of smallest eddy in an isotropic turbulence, where ε is the rate of energy dissipation per unit mass of gas. Associated with this length scale is a time, τ, characteristic of the lifetime of the eddy

$$\tau = (\nu/\varepsilon)^{\frac{1}{2}} \tag{41}$$

There are also two corresponding properties of the particle; its diameter or characteristic dimension d_p, and its "relaxation time" τ_p characterising the time it takes
to conform to changes in gas velocity. For small *relative* velocities between
particle and gas, Stokes Law can be used, and τ_p is given (for a sphere) by

$$\tau_p = \frac{d_p^2 \, \rho_p}{18 \mu} \qquad (42)$$

Provided first that the particles can fit into the smallest eddy, i.e. $d \ll \lambda$,
and secondly that they can follow the smallest eddy's motion, i.e. $\tau_p \ll \tau$, then
we can use an average shearing rate in the eddies

$$\overline{\frac{dV}{dx}} = \left(\frac{\varepsilon}{15\nu}\right)^{\frac{1}{2}} \qquad (43)$$

to calculate the collision rate. This approach has been discussed by Delichatsios
and Probstein (ref. 59) and Abrahamson (ref. 60), and has been found by Okuyama,
(Ref. 61) to accurately describe agglomeration rate of 1.5 μm dia. particles in
an 8 m s^{-1} air flow. The collision rate N per unit volume (two-body collisions)
is naturally proportional to the square of particle number concentration n_p and is
given by

$$N = 1.3 \; d_p^3 \left(\frac{\varepsilon_o}{\nu}\right)^{\frac{1}{2}} n_p^2 \qquad (44)$$

For example, under the conditions of Okuyama, Kousaka and Yoshida (Re_{pipe} =
1.5×10^4, 26 mm dia., particle density 1400 kg m^{-3}) using the average energy
dissipation rate throughout the pipe, given by

$$\varepsilon_o = f\overline{V}_z^3/D \qquad (45)$$

where f is the friction factor, ε_o = 170 W kg^{-1} of air, and for a particle mass
concentration of 10 g m^{-3}, each particle will collide with another on average 0.06
s^{-1}, or once every 130 m. Thus for many industrial situations, turbulent agglomeration of particles around 1 μm with each other is expected to be negligible.

However, for larger particles, the effect of their inertial cannot be overlooked,
and Abrahamson (ref. 60) has calculated a collision rate for particles catapulted
from eddy to eddy (Fig. 10(b)), so that the random approach and velocities of the
gas kinetic theory could be used. In doing this he has also used the existence
of the inertial subrange of eddies (larger eddies unaffected by viscosity) at
high Reynolds number. His rate is

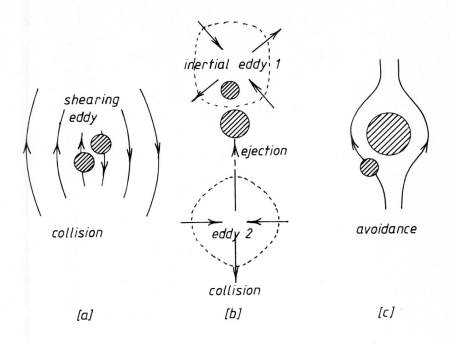

Fig. 10. Mechanisms of particle collision in a turbulent gas (a) Collision in the shear field of the smallest eddy (b) Collision by catapult from neighbouring eddies (c) The close approach (separation $<d_p$) problem.

$$N = \frac{7.1 \quad d_p^2 \; V' \; n_p^2}{\left[1 + 1.5 \; \tau_p \varepsilon_o / (V')^2 \right]^{\frac{1}{2}}} \qquad (46)$$

$\quad$ (V' is the gas fluctuation velocity)

Thus for a mass density of 10 g m^{-3} of larger 10 μm dia. particles flowing under the same conditions as above (the *number* density is much less, $V' \approx 0.03\overline{V}$) one calculates a collision rate for each particle with another of about 1 per second, or 1 every 10 m of pipe. It can be seen that in many industrial ducts, significant agglomeration may occur before reaching the cyclone. If very fine particles have reached 10 μm size by agglomeration during the factory process, then they too will suffer this enhanced agglomeration. Williams and Crane (ref. 62) have applied Abrahamson's theory to agglomeration in a wet steam pipe, and note the much higher agglomeration rates close to the pipe wall, where ε is much higher than the average ε_o. If the dust is distributed in size, as is most common, then the enhanced collision rate expected from eqn. 45 will also apply to smaller particles of 1 μm dia. or less, provided their collision partner is of the order 10 μm or more. Abrahamson (ref. 60) has given the limits of availability of eqn. 45

The above calculations have ignored the "close approach" problem of premature interaction and avoidance of particles on a collision course, through the intervening gas (see Fig. 10(c)). The particle inertia is overriding and the "collision efficiency" close to 1, however, when the Stokes number

$$\psi = \frac{\rho_p \, V_{rel} \, d_{p1}^2}{18 \, \mu \, d_{p2}} \tag{47}$$

(defined with V_{rel} as the relative velocity between a particle 1 and larger particle 2) is 1 or greater (see Soo, ref. 63). This is doubtful for most collisions of Ikuyama, Kousaka and Yoshida's study (1.4 μm dia. = d_{p1} = d_{p2}, use $V_{rel} = \lambda/\tau$ as a maximum so that ψ max = 0.7), whereas it is a better assumption for 10 μm particles (average $\psi \approx 3$ for the above example).

3.3 Wall deposition

Deposition of particles in turbulent flow involves penetration of the laminar sublayer by the particles with subsequent sticking if the approach velocity is not too high. Dahneke (ref. 64) gives limiting velocities above which bouncing will occur, and these are 1.2 m s^{-1} for easily deformable 1.3 μm spheres (latex on quart) down to 0.1 m s^{-1} for more rigid 1 μm spheres (quartz on quartz). Cleaver and Yates (ref. 65) have pointed out the importance of the intermittent powerful "downbursts" of gas towards the wall (found recently in turbulent boundary layers) in carrying particles very close to the wall. Their models successfully describe a wide range of measurements. Deposition rates increase rapidly with particle size from 1 μm until a size 10-20 μm is reached, above which a maximum deposition flux is reached. This maximum, in dimensionless form, is

$$N/CU^* = 5 \tag{48}$$

where N is the flux of particles (number) per unit area of wall, C is the number concentration in the turbulent core, and U* is the friction velocity of the shear flow at the wall, defined by the wall shear stress τ_w.

$$U^* = (\tau_w/\rho)^{\frac{1}{2}} \tag{49}$$

Evaluating this maximum flux for the 10 g m^{-3} flow of 10 μm particles mentioned above (we will take 8 m s^{-1} in a 0.15 m dia. tube, sheet metal surface roughness 120 μm) we find a particle will deposit on average 7 times per metre of tube length. This rather astounding figure does not account for limitations of turbulent diffusion from the central core to where the boundary layer mechanism operates. El-Shobokshy and Ismail (ref. 66) calculate similar deposition velocities

N/C (ours is 2 m s^{-1}) at high tube Reynolds numbers. They show that greater
roughness of the wall increases the deposition rate of smaller particles (0.1-5 µm)
but does not influence larger sizes.

3.4   Resuspension of particles from a wall

Cleaver and Yates (ref. 67) have discussed the rather meagre experimental work
in this area, and have noted that particles are not seen to roll once stuck on
the surface, and begin to be removed only after the main flow has become turbulent.
Cleaver and Yates argue that resuspension depends on lifting forces from the intermittent bursts and derive an expression for the fractional rate of particle
removal dR/dt which is dependent solely on the overall wall shear properties.

$$\frac{dR}{dt} = -2 \times 10^{-7} \, \tau_w/\mu \qquad\qquad\qquad (50)$$

This relationship described quite well the removal of particles of the order
1 µm in size. If we apply it to our pipe flow example ($\tau_w = 0.9$ N m^{-2}) above,
we find a resuspension rate of 0.01 s^{-1}, a value which is negligibly small compared
with other rates.

3.5   Radial dust movements in a pipe

The large number of collisions expected with the wall (from section 3.3)
obviously does not result in all the dust remaining there, but on average, particles
appear to spend a disproportionately large part of their time close to the wall.
Arundel, Bibb and Boothroyd (ref. 68) show this especially for large concentrations
(3.6 kg m^{-3}) of 0-40 µm zinc particles flowing vertically in a 75 mm diameter
tube. They show higher concentrations towards the wall, and also that this
maldistribution increases with higher velocity. They attribute this behaviour to
electrostatic charging and mutual repulsion, agglomeration near the wall, and a
damping of turbulent diffusivity near the wall because of the inertia of the
particles. Boothroyd and Walton (ref. 69) show evidence from gas tracer dispersion
that for the same system, gas diffusion was markedly reduced close to the walls,
the gas shear rate also was reduced, and the shear stress was being carried largely
by solid particle impacts. Zisselmar and Molerus (ref. 70) have made detailed
measurements of *gas* fluctuating velocities in suspensions of 53 µm particles, with
laser doppler anemometry. They find that concentrations as low as 30 g m^{-3} make a
significant difference to the fluctuating velocities, especially near the high
shear layer (~200 µm from the wall). The longitudinal velocities are reduced, as
expected, but the radial fluctuations are *increased* , probably by particle bounce.
It is significant that the radial velocity fluctuations continue in towards the
wall (to within 50 µm) at the same value as at 1-2 mm, as if the wall was not

present. If essentially all of this turbulence was caused by the particles, and they were reflected off the wall with little momentum loss, one can understand this behaviour.

In horizontal pipes, gravity is expected to cause maldistributions in longer pipes (e.g. a 50 μm dia. particle will fall about 0.3 m in 2s) and this will be enhanced by agglomeration.

From the above discussion, it is doubtful whether many fine dust flows fed into cyclones are uniformly dispersed and in an unagglomerated state, as is commonly supposed.

3.6 Dust motion on entering a cyclone

There is some justification for ignoring the gas radial velocity in the first turn the gas makes on entering a cyclone, as in all practical designs of return-flow cyclone, the exhaust extends to at least the depth of the inlet. Intensive analytical modelling has been done of particle paths in typical cyclone vortex flows with no radial inflow, perhaps to suit the first turn. Strauss (49) shows that such a calculation predicts collection of particles as small as 3 μm on the wall of a 500 mm dia. Stairmand cyclone with V_i = 15 m s^{-1} before the first complete rotation. Maslov, Lebedev, Zverev and Ushakov (ref. 71) have constructed ducts with complete 360° turns and constant cross-sectional area, to measure *gas* velocities, and use these to test the normal assumptions made in calculating dust trajectories. They found experimentally that 16 μm dust (apparently well dispersed) fed in a dilute jet and entering the curved duct at 17.5 m s^{-1}, was collected on the wall in 0.3-0.5 of a turn (outside diameter 0.5 m-2 m). They could predict accurately the path of the particles provided that aerodynamic equivalent diameters were used, and radial gas flows were known. Turbulent influences on particle drag and dispersion could be ignored in calculating the *mean* path (turbulent particle diffusion, however, was certainly noted.) It **thus appears,** that when the dust is fully dispersed **into its** primary particles, and there is negligible radial inflow, collection at the wall should be rapid and soon after entry to a normal cyclone.

In the more common situation where particle concentrations are high enough for agglomeration to have occurred before entry, collection at the wall is expected to be even more rapid. Reznik and Matsnev (ref 44) show dust concentrations C_p measured within the first revolution of gas in a 254 mm cyclone (peat dust 27 μm). Compared to an average inlet concentration of 21.6 g m^{-3}, C_p varied logarithmically, from 80 g m^{-3} near the outside wall down to 0.02 g m^{-3} near the outer wall of the exhaust duct. However, below the exhaust lip, Reznik and Matsnev show (as given in Fig. 11) that the central concentration has risen again (to 0.2 g m^{-3}). This is the concentration that is emitted, so we have apparently lost the initial gain in separation.

Reznik and Matsnev have also shown by considering dust concentrations and the inward radial flow just beneath the exhaust lip (described in 2.4.2 as coming from slower-rotating boundary flow) that 85 per cent of the dust lost to the exhaust in their cyclones was convected inwards with this leak flow. As the inward velocities were in the range 5-8 m s^{-1}, only large agglomerates would not have been convected inwards. It seems that this leak flow is probably the reason why lower efficiency is often associated with lower pressure drop.

3.7 Turbulent diffusion of particles in a cyclone

We have seen above that large concentrations of dust accumulate soon after entry to the cyclone close to the walls, leaving large concentration gradients acting to diffuse particles towards the centre. Hejma (ref. 40) similarly found an order of magnitude drop in dust concentration over a distance 20 mm from the wall, with a polydisperse dust of mean size 27 μm. In contrast, with a fine (1-2 μm) monodisperse dust, concentrations along the radius were observed to be much more uniform. Hejma understandably attributed this difference to turbulent diffusion, as he measured by hot wire anemometers radial turbulent velocities V'$_r$ of 2 m s^{-1} over much of his cyclone, rising to 5 m s^{-1} in the cone (he compared this with the calculated radial drift velocity 0.135 m s^{-1} of a 3 μm particle in his vortex flow). There are difficulties, however, with this interpretation, for the dust concentration in the upper cyclone had a concentration *maximum* in the centre.

To assess the importance of diffusion, we need first to decide whether the turbulence is characteristic of the wall shear layer, or of the moving vortex in centre, and secondly to allow for the inertia of particles in reducing their effective diffusivity. We will calculate quantities for the cyclone used by Hejma (D = 314 mm, core length 5D, v$_i$ = 17 m s^{-1}). First we can treat the cyclone as a pipe with axial flow, only we will use as a characteristic velocity V$_\theta$ in the region close to the wall but outside the boundary layer (of about 12 mm, see Hejma). This was close to 20 m s^{-1}. Mizushina and Ogino (ref. 72) have shown that for pipe Re > 10^4, eddy viscosities ν$_T$ are constant over the core of the flow and are given by

$$\frac{\nu_T}{\nu R^+} = 0.07 \tag{51}$$

where R$^+$ is the dimensionless pipe radius, given in terms of the friction velocity at the wall U* (eqn. 49) by

$$R^+ = RU*/\nu \tag{52}$$

For a wall roughness of 120 μm, U* = 1.0 m s^{-1}, giving ν$_T$ = 0.011 m^2 s^{-1} for

R taken as the cyclone radius. If we note the similarity of turbulent momentum and mass transfer (e.g. Bird, Stewart and Lightfoot ref. 73) the ratio of ν_T to the gas diffusivity $\mathcal{D}_T$ is found experimentally to be 0.7 to 0.9 in conduits, and 0.5 in jets. We will use

$$\mathcal{D}_T = \nu_T/0.8 \tag{53}$$

and thus calculate $\mathcal{D}_T = 0.014 \ m^2 \ s^{-1}$. This compares with values of eddy diffusivity in pipes given by Boothroyd (ref. 74)

$$\frac{\mathcal{D}_T}{\overline{U}_{pipe} \ D_{pipe}} = 1 \ to \ 2 \times 10^{-3} \tag{54}$$

for $10^4 < Re < 10^6$. This gives $D = 0.006 - 0.012 \ m^2 s^{-1}$.

An approach which is more fundamental, and can give a better insight into the diffusion mechanism, is that given by Weinstock (ref. 75), who calculates for homogeneous, isotropic turbulence via analytical approximations for the Lagrangian decay time, that

$$\mathcal{D}_T = \left(\frac{m-1}{m}\right) \frac{V'_o \ L_o}{4\pi^{\frac{1}{2}}} \tag{55}$$

where m is the wavenumber power index in the turbulent energy spectrum, v'_o is the root-mean-square fluctuating velocity, and L_o, is the outer scale length of the turbulence. In pipe flow, L_o, will be taken as πR (Brodkey (ref. 76)) and generally $V'_o \simeq 0.035 \ \overline{U}$ pipe. We will assume that an inertial subrange dominates the diffusion since the Re are large, so $m = 5/3$. Thus for an equivalent pipe, we obtain $\mathcal{D}_T = 0.019 \ m^2 \ s^{-1}$, which is not too much different from the other estimates.

Eqn. 55 is useful, because we can use the measured values of V'_o, in Hejma's cyclone, with an appropriate L_o, to calculate $\mathcal{D}_T$. Hejma found larger $L'_o/\overline{V}$ in his cyclone (0.06 to 0.14) than in pipe flow (0.035) with $V'_r \approx V'_z$, i.e. isotropic turbulence, out from the wall. Turbulent intensities $V'_o/\overline{V}$ increased both at the wall and towards the centre, showing turbulence generation in both these regions. Using Hejma's value of $V'_o = 2 \ m \ s^{-1}$, and πR as L_o, we calculate $\mathcal{D}_T = 0.056 \ m^2 \ s^{-1}$. Alternatively we might take π times the diameter of the vortex $(0.35 \ D_i)$ as a characteristic L_o, in which case we obtain $\mathcal{D}_T = 0.030 \ m^2 \ s^{-1}$ *

---

* Using $\nu_T = 0.8 \mathcal{D}_T = 0.024 - 0.045 \ m^2 \ s^{-1}$, we can evaluate the characteristic radial velocity given in eqn. 28. $V_R = AR = 2 \nu_T C/R_i = 4.0 \ to \ 7.4 \ m \ s^{-1}$ for Hejma's cyclone. Perhaps a more meaningful radial velocity would be that at $R_b = 0.35R$ (for $C = 10$), $V_{Rb} = AR_b = 1.4 \ to \ 2.6 \ m \ s^{-1}$, which is also in the range of Hejma's measured V_r. For cyclones, the value of Λ in eqn. 24 then falls in the range 1.7 to 3.2. This is almost an order of magnitude larger than that from unconfined trailing vortices (see 2.2).

These diffusivities need to be corrected for the fact that larger particles do not follow the gas completely. Yuu, Yasukouchi, Hirosawa and Jotaki (ref. 77) have found a linear relationship for $\mathcal{D}/\mathcal{D}_p$ versus a Stokes number ψ, from measurements in the steady state region of a dust-laden air jet. Their Stokes number can be expressed as

$$\psi = \tau_p/T \tag{56}$$

where T is a jet characteristic time (diameter/average velocity), and their relationship was

$$\frac{\mathcal{D}_T}{\mathcal{D}_p} = 1 + \psi/12 \tag{57}$$

($\mathcal{D}_p$ is the particle diffusivity). If we rather arbitrarily take $T = R/V_\theta$ in a cyclone, we find in Hejma's cyclone $T = 7.8$ ms. For his fine dust (1.5 µm) ψ is then 10^{-3}, and $\mathcal{D}_p = \mathcal{D}_T$, whereas for his course dust (27 µm), $= 0.3$, so that $\mathcal{D}_p \approx 0.97 \mathcal{D}_T$. For practical purposes we can take the turbulent particle diffusivities as equal to turbulent gas diffusivities for most particles of interest.

It is now of interest to calculate the mean square displacement Δr in the radial direction due to diffusion. For Hejma's cyclone, within 40 mm of the wall, the axial velocity >2 m s^{-1}, giving a residence time t of <0.4s in the cylindrical portion. Treating the problem as a plane diffusion,

$$\Delta r = (2 \mathcal{D}_T t)^{\frac{1}{2}} \tag{58}$$

so that Δr has values < 0.15 m to 0.2 m for $\mathcal{D}_T = 0.03$ to 0.056 m^2 s^{-1}. This is expected to be the standard deviation of a Normal distribution of radial migrations (see Abrahamson, Martin and Wong, ref. 43), which is opposed by outward migration due to centrifugal action and aided by inward migration due to gas convection. Using the conditions at 40 mm from the wall, the distance moved radially under the centrifugal action is 0.02 m s^{-1} x 0.4 s = 0.009 m, and 3.0 m, for Hejma's fine and coarse ducts respectively. It is apparent that diffusion overides the centrifugal effect for the 1.5 µm dust (treated as primary particles), but is negligible for the 27 µm dust. For the larger dust, one expects that the concentration profile seen at the wall will be the result of a steady-state balance between the outward centrifugal flux of particles, and the inward diffusional flux down the concentration gradient. Thus

$$n_p \, Vr_p = + \mathcal{D}_T \frac{n_p}{\partial r} \tag{59}$$

where n_p is the number concentration of particles and V_{rp} is the radial centrifugal velocity of the particles. Rearranging eqn. 59.

$$\frac{\partial \ln n_p}{\partial r} = + \frac{V_{rp}}{\mathcal{D}_{\mathsf{T}}}$$

(60)

Since the mass concentration of particles $C_p = \rho_p n_p$ it does not matter if we use C_p or n_p in eqn. 60. For the 27 μm dust, the experimental value of $\partial \ln C_p / \partial r$ (the plot is linear) was −170, compared with −137 to −257 for $V_{rp}/\mathcal{D}_{\mathsf{T}}$, using the above two estimates of $\mathcal{D}_{\mathsf{T}}$. This agreement is confirmation of the mechanism, and under-lines the adequacy of using the primary particle size for larger sized dusts.

For the fine dust, however, well down into the cone there is also a rapid rise of concentration close to the wall in Hejma's cyclone. $\partial \ln C_p / \partial r$ is −160 for the first 8 mm out from the wall, and evaluating a smaller $\mathcal{D}_{\mathsf{T}}$ from eqn. 51, leads to values of V_{rp} of 4.6 m s^{-1} (from eqn. 60 . With the much higher centrifugal acceleration, this corresponds to an effective particle diameter (sphere, same density as the primary particles) of 9 μm. Since it is expected that agglomerates have been formed, with perhaps 0.3 the particle density, we expect agglomerates of about 15 μm in size.

This success in describing dust concentrations by diffusion does not however apply towards the centre of the cyclone, where concentrations have been found to have a maximum (large for fine dust, Hejma, and at much lower concentration for coarse dust (Reznik and Matsnev ref. 44 see Fig. 11).

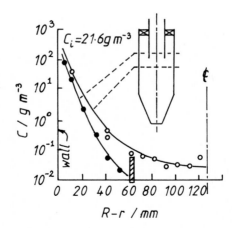

Fig. 11. Concentration distribution of dust(mean size 27 μm) in a vane-entry cyclone.

3.8 The dust rope and collection on the wall

The Görtler vortices described in 2.4.5 appear to cause bands of alternating
slow and fast gas flow down the wall of a cyclone. The gas in regions with higher
V_θ will move towards the wall, whereas those with lower V_θ will move radially
inward, in a steady exchange, in much the same fashion as with instabilities
above a heated surface in a gravitational field. This steady wall vortex movement is expected to concentrate dust in the fast gas flow regions, where as well
as gas instability, we now have dust concentration differences, enhancing the
instability by effective fluid *density* increases. It is reasonable that most of
the dust in cyclones fed at higher concentrations will move down the wall in
distinct bands, as often observed (e.g. Muschelknautz ref. 79).

The density instability has more implications in a cyclone than one might suppose.
One does not need *agglomeration* of fine particles - i.e. the attachment of them
to each other by surface forces etc - to have them move under the centrifugal
acceleration faster than they would when isolated particles. One merely needs
them to come into close proximity with each other, and fluid forces can act to keep
them together in a stable fast-settling swarm. Adachi, Kiriyama and Yoshioka ref.
80 have given both theoretical and experimental work on the settling velocities
(under a gravitational field in their case) of swarms of particles which carry the
intervening fluid with them, and settle more as a solid body. For laminar flow,
and low swarm solid concentration C_p(<0.001 mass solid/mass fluid) they found the
terminal velocity U_s given by

$$U_s = \frac{C_p D_s^2 g}{15\nu} \qquad (61)$$

where D_s is the swarm diameter (eqn. 61 has assumed $\rho_p \gg \rho_p$). We can compare
U_s with the terminal velocity V_t of isolated particles given by Stokes' Law

$$V_t = \frac{d_p^2 \rho_p g}{18\mu} \qquad (62)$$

If $U_s > V_t$ the swarm will remain together and increase the sedimentation velocity
(whether under gravity or in the cyclone). That is, if the swarm effect is to be
important, from eqns. 61 and 62,

$$\frac{D_s}{d_p} < \frac{6}{5} n \qquad (63)$$

where n is the number of particles in a swarm. This number will depend on the
frequency of close approach of particles, which is expected to be proportional to
the square of the number concentration of particles, as in Abrahamson's turbulent

collision theory (ref. 60). Eqn. 63 is not too severe a condition, for it predicts that a linear chain of particles very close to each other but not held by any particle-particle forces, should not separate. In practice, eqn. 61 only applies to a spherical swarm.

The particle concentration at which settling swarms become important is likely to be high, and to be achieved only towards the wall or in the bin of a cyclone; Yerushalmi, Turner and Squires (ref. 81) have noticed that under gravity, with a 3-5 m s^{-1} vertical turbulent flow, a solids concentration of about 1 kg m^{-3} was necessary before segregation into dense and dilute regions was noticed by high speed photography. In their "fast fluidised bed", the slip velocities of solids relative to the gas was an order of magnitude larger than the terminal velocities of individual 60 μm particles. Their experiments could not detect small swarms or agglomerates, however, and it is these which will be important for more severe conditions in a cyclone. Once a stable swarm (or stable agglomerate) is formed, it will overtake most primary particles in its path to the wall, and it is expected that some of these smaller particles will be collected by the larger entity. Some more formal work on the stability of solid-fluid suspensions is beginning to be done (e.g. Hill and Bedford ref. 82) but this work has not yet been applied to experiment.

In order to give some more direct evidence of the condition of dust close to the wall of a cyclone, I have used a powerful 1 μs flash lamp to photograph individual particles in the cyclone of Abrahamson, Martin and Wong (ref. 43). If the flash lamp (detailed in Appendix I) was used in back lighting through a window from the other side of the cyclone in conjunction with a 35 mm camera and a 0.7 m extension tube, one could define particles as small as 5 μm moving at velocities up to 30 m s^{-1}. Fig. 12 shows some single exposures and some **double** exposures (with 10 μs separation) of the **alumina-cryolite** dust detailed in appendix II (50 per cent < 15 μm). **The** feed, dispersed at 30 m s^{-1}, is shown in Fig. 12(a) with the largest particle at 50 μm. For the remainder of the photographs, the jet ejector velocity was increased to greater than 100 m s^{-1} so that an adequate dispersion was probably attained, feeding the cyclone with an inlet concentration of 10 g m^{-3}. Fig. 12 (b), (c) taken through a glass window 400 mm up the cone from the dust exit, as the dust rope moved over it. The window was provided with small inward air leaks which kept the edges clear of piling dust, but did not disturb the viewed area. In contrast, 12(d) shows the window region much less populated in the absence of the dust rope (this swept past the window several times a second). The dust particles moved downwards at 7 m s^{-1} and about 15°, irrespective of whether the dust rope was present or not. There is no evidence of agglomerates, and many of the particles seen are about 20 μm in size. When one focusses in from the window however, there is some evidence of agglomerates. Fig. 12 (e) shows a

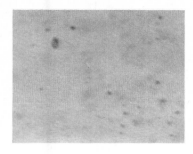

(a) Feed ejector exhaust, double with 5 μs delay, 30 m s^{-1}.

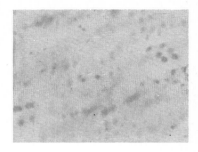

(b) At window of cyclone, double with 10 μs delay, 7 m s^{-1}.

(c) At window, double with 10 μs delay, 8 m s^{-1}.

(d) At window, double with 10 μs delay, 7.4 m s^{-1}.

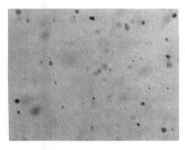

(e) 2 mm in from window, single exposure

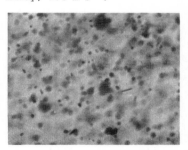

(f) 2 mm in from window, double with 10 μs delay, largest particles 70-80 m.

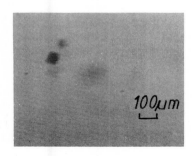

100μm

(g) 10 mm in from window of cyclone, single exposure.

Fig. 12 Photographs of alumina-cryolite dust with 1 μs flash. Scale: 20 μm to the mm.

single exposure 2 mm in, indicating many particles 5-20 μm in size, probably primary particles whereas 12(f) shows the dust rope, with much higher local concentration, with particles moving at about 13 m s^{-1}. Fig. 12(f) is a double exposure, showing several large particles (60-100 μm) which are not seen moving along with the smaller particles. When one focusses further into the cyclone, the number of particles becomes very small, with sizes from 5 μm, but again larger particles are seen - e.g. those of the order 50-100 μm in Fig. 12(g), taken 10 mm in from the window. At 20 mm in, most photographs did not show particles at all.

The major conclusion to be drawn from these photographs of a coarse dust is that very few agglomerates exist close to a cyclone wall (at least with a polydisperse dust > 5 μm). This is **contrary** to expectations from calculations of collision rates in the highly concentrated dust layer near to the wall. The answer appears to lie in the efficiency of collisions with the wall in breaking up agglomerates. For an average dust particle of 15 μm dia., density 3000 kg m^{-3}, the radial velocity onto the wall is expected to be (Stokes' Law) 3.5 m s^{-1}, which is probably sufficient to break up any agglomerates of these larger primary particles (see Kousaka, Okuyama, Shimizu and Yoshida (ref. 56).). The larger dust entities seen in the photographs are likely to be agglomerates which have arrived from the core of the cyclone. They are likely to have radial velocities of 20 m s^{-1} or more, explaining why they were only seen once in the double exposure.

3.9 Recirculation of dust and bin collection

In the photographic study above, I noticed that the dust ropes moving past the window remained when the dust feed to the cyclone was switched off. Only when the bin was emptied of dust, were the dust ropes removed. This is evidence for recirculation of dust.

In 2.4.1 we discussed the gas circulation into the bin and out again along the centreline, amounting to 10-15 per cent of the total gas flow. Abrahamson, Martin and Wong (ref. 43) have measured velocities in the bin, and found the circulation in and out of the bin sufficient to prevent settling of particles less than 150 μm, before they were carried out into the cyclone again. If one considered a centrifugal deposition onto the dust heap, particles of size less than 50 μm would be returned to the cyclone. The collected particles were definitely *convected* into the bin, for tests done with inverted cyclones showed very little reduction in collection efficiency. Abrahamson, Martin and Wong concluded that agglomeration was necessary for the deposition of most dusts onto the dust heap, and the agglomerates needed to be at least 50 μm in size.

In view of the wall disruption of agglomerates noted above, the agglomeration will have to take place as the dust leaves the cyclone and is showered into the bin. The dust concentrations in the bin are high (the pick-up from the vortex end which plays about on the dust heap ensures this) and dust has about half its cyclone collection time in the bin. We have no direct evidence for the dust behaviour in a bin, but it is reasonable that the bin can be regarded as an "agglomerator". Large bins, which collect dust with somewhat higher efficiencies, have low tangential gas velocities close to their walls (2 m s^{-1}, for a cyclone wall velocity of 14 m s^{-1}). Thus break-up of agglomerates should not be a problem in these large bins.

Small leakages of air into bins are known to cause large increases in emission from the cyclone. Tests done by Abrahamson, Martin and Wong on cement dust elutriation from already-collected dust in the bin showed no greater loss rates when 5 per cent of the total flow was leaked into the bin. They also noticed that leaks of this size were sufficient to prevent the "dust cloud" from entering the bin. The leak appears to satisfy the vortex axial flow without it having to enter the bin. These observations contradict the commonly held view that "base pick-up of dust" from the bin is wholly derived from dust which has been deposited on the heap, and support the recirculation idea.

Recirculation from the bin back into the cyclone may determine the overall efficiency of the cyclone, depending on its size and design, and the dust characteristics. Hejma (ref. 40) showed a dust concentration in his cyclone which was *higher* in the centre than towards the walls, for his fine 1.5 µm dust. This can have come only from the bin, in the rapidly ascending central core. Reznik and Matsnev (ref. 44) observed in a transparent cyclone 254 mm in diameter, "periodic motion of dust filaments in the volume of the cyclone". They reported that "spirals of air suspension now accumulate dust, now discard it into the bin".

The crucial dust concentration is that just below the gas exhaust, as this determines the emitted dust. We can now conveniently divide the behaviour of dusts in a cyclone between that of coarse dust (> 10 µm) and that of fine dust (0.5 - 10 µm).

3.10 Coarse dust

Particles greater than about 10 µm, after being recirculated from the bin, are expected to be largely recaptured near the wall again rather than reach the gas exit, provided the distance between the dust and gas exits is sufficiently large. Ter linden (ref. 18) effectively showed this by his determination of efficiency when a coal dust (normal collection 95 per cent) was fed into the cyclone from various points within the cyclone. Provided that the dust was fed no higher than half way towards the gas exit from the dust exit, the efficiency remained 95 per cent.

In the case of an adequately long cyclone, then, where bin-cone recirculation effects have been removed from the gas exit area, the steady state concentration profile resulting from a balance between the radial particle velocity and back-diffusion eqn. 60 will apply. To a first approximation we can ignore agglomeration within the core of the cyclone because concentrations are so low, and also immediately next to the wall as discussed above, but perhaps not so easily just outside the wall region. The radial *time averaged* particle velocity V_{rp} is the sum of the gas velocity V_r and the velocity relative to the gas $\tau_p V_\theta^2/r$

Thus

$$V_{rp} = \tau_p \frac{V_\theta^2}{r} + V_r \tag{64}$$

where V_r is normally negative (inward). Using the Burger's vortex form

$$V_\theta = \frac{V_i R_i}{r} \left(1 - \exp\left[-C(r/R_i)^2 \right] \right) \tag{65}$$

related to the inlet velocity V_i and average inlet radius R_i, we have the following equation governing the concentration profile of the jth size of particle

$$D_T \frac{d\ln C_j}{dr} = \frac{\rho_p d_p^2 V_i^2 R_i^2}{18 \mu r^3} \left(1 - \exp\left[-C(r/R_i)^2 \right] \right)^2 + V_r \tag{66}$$

where Stokes' Law is assumed, C has a preferred value of around 10, and V_r varies from zero at the wall, to V_{rb} at the position of maximum V_θ. The characteristic radial velocity V_{rb} may be calculated from the assumed constancy of the turbulent radial Reynolds number C. If we assume an unbounded vortex, we can use the Reynolds number to the 0.5 power relationship eqn. 24, found suitable to describe ν_T for trailing vortices. This leads to

$$V_{rb} = -0.35 \ 2^{3/2} \ \pi^{\frac{1}{2}} \ C\Lambda \left(\frac{V_i \nu}{R_i} \right)^{\frac{1}{2}} \tag{67}$$

where Λ may be taken as 2.5 (see discussion on eqn. 55

On the other hand, if we take $\nu_T = 0.8 \ D_T$ as in eqn. 53 and D_T is related to size of the bounding walls as in Weinstock's relation eqn. 55 using $V'_o = 0.1 \ V_i$ as in Hejma's cyclone, then we find

$$V_{rb} = -0.032 \ C^{\frac{1}{2}} V_i = -0.10 \ V_i \tag{68}$$

where a value of 10 has been used for C. It can be noted that this estimate of V_{rb} follows the customary assumption with cyclones, that $V_{rb} \propto V_i$. Also the magnitude of V_{rb} is seen to be close to that of the mean fluctuating velocity V'_o.

Expression 68 relates to the boundary dimension (diameter) of the cyclone through a scale of the turbulence, but does not consider the length of the vortex. In this sense the model assumes that the length of the cyclone is just that which allows the total gas flow to turn inward at the rate dictated by the turbulent radial transport of angular momentum (in practice this would be the optimum length). The equation 66 may be solved numerically, given that the average concentration across any cross section is approximately the inlet concentration C_i. The efficiency η_j is given by

$$\eta_j = 1 - \frac{\int_0^{P_e} C_j \bar{V}_e 2\pi r dr}{C_i V_i A_i} \tag{69}$$

If we take the diffusivity $\mathcal{D}_T$ from eqn. 55 with m = 5/3, and assume a linear variation of V_r between the wall and V_b we will have enough information to find a solution. The nomenclature is made clearer in Fig. 13. The constants used in eqns. 67 or 68 for V_{rb} may have to be adjusted so that the total inflow equals the

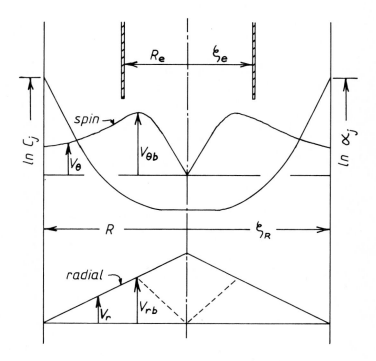

Fig. 13. Steady state profiles for the coarse dust model.

total gas flow (or 0.9 of it), and we may have to make sure that the cyclone we are dealing with has negligible inflow just below the exhaust lip.

The solution of eqn.66 is beyond the scope of this chapter, but by making it dimensionless, we can see what numbers will be important for scaling cyclones. If we take some fraction of V_{rb} as a characteristic V_r and take

$$V'_o = 0.10 \ V_i \quad \text{(Hejma)} \tag{70}$$

$$L_o = \pi D_i$$

in expression 55 for $\mathbf{D_T}$, we obtain

$$\frac{d \ln C_j}{d\xi} = 3.2 \ \text{Stk}_i \ (1 - \exp[-c\xi^2])^2 - \begin{cases} 150 \ \dfrac{(\xi_R - \xi)}{\xi_R - \xi_b} \ C \ \Lambda \ \text{Re}_i^{-\frac{1}{2}} & \text{(unbounded)} \\[4mm] \text{or} \ 1.8C^{\frac{1}{2}} \ \dfrac{(\xi_R - \xi)}{\xi_R - \xi_b} & \text{(bounded)} \end{cases} \tag{71}$$

where $\xi = r/R_i$,

$$\text{Stk}_i = \frac{\rho_p \ d_p^2 \ V_i}{\mu R_i}$$

$$\text{Re}_i = \frac{R_i V_i}{\nu}$$

the constants C, Λ (and therefore ξ_b) are weak functions of Re_i, and the numerical constants are the best estimates only.

The efficiency is then given by

$$\eta_j = 1 - \frac{2}{\xi_e^2} \int_o^{\xi_e} \alpha_j \ \xi \ d\xi \tag{72}$$

where $\alpha_j = C_j/C_i$, $\xi_e = R_e/R_i$, $\xi_b = r_b/R_i$, and $\xi_R = R/R_i$.

It can be seen that the efficiency depends on four dimensionless groups; Stk_i, Re_i, ξ_R and ξ_e. The concentration at the exhaust will diminish, and so the efficiency will increase, with increase in Stk_i, (and for the unbounded case) with a decrease in Re_i. This trend corresponds to those of the empirical $\text{Stk} \ \text{Re}^{-0.1}$ or $\text{Stk} \ \text{Re}^{-0.25}$ criteria mentioned in 1.2. Thus the experimental findings support the unbounded case, or perhaps some intermediate Re_i dependence between the bounded and unbounded extremes.

3.11 Fine dust

We have ignored agglomeration in the separation of coarse dust but we cannot do so with fine dust of less than 10 μm in size. From our calculations of diffusion, these finer primary particles should be collected very little, and indeed in many cyclones this is so. However some cyclones collect most of the dust down to 1.5-2 μm, and we have seen in 3.7 that agglomerates of 15 μm size were necessary to explain the rapid rise of dust concentration (1.5 μm dust) towards the conical wall of Hejma's cyclone. His overall efficiency for this dust was 40 per cent. It appears that the controlling step for efficiency has moved away from the exhaust mouth, to the conical or lower part of the cyclone, when fine dust is to be collected. Avant, Parnell and Sorenson (ref. 22) systematically varied both the barrel and cone lengths with cyclones collecting fine sorghum dust. They found a highly significant decrease in the cut size (6 to 3 μm) for an increase in the cone length, but not for an increase in barrel length. The dust concentration measurements of Hejma showed 12 g m^{-3} in the downward moving gas close to the conical wall, near the dust exit, and the same concentration at the centreline, in the upward flow of gas from the dust bin, with a lower concentration between. It is reasonable to conclude that the cyclone cone acted as a concentrator (the concentration fed to Hejma's cyclone was 3 g m^{-3}), that the concentration in the bin was 12 g m^{-3}, and that some of this bin suspension was recirculated to the cone. If all the dust ultimately collected (40 per cent of 3 g m^{-3}) was concentrated into the 10 per cent bin flow, we would have 12 g m^{-3} entering the bin. It appears that the centrifugal forces in the bin do not collect all of the agglomerates on to the dust heap, and those not collected have a chance of being thrown to the wall on reentering the cyclone. Addition to, or subtraction of dust from the bin-cone recycle may be regarded as a controlling step, and is taken here as   the overcoming of diffusion within the cone by the formation of stable agglomerates or particle clusters which have high terminal velocities.

Particle clusters and agglomerates depend for their formation first on the frequency of "near approaches" of primary particles, which is proportional to the square of the particle concentration, as outlined above. Secondly, the character of the close encounter is important, including the relative velocity of particle-particle approach through the viscous intervening gas, and any longer range particle-particle forces (electrostatic). Thirdly the agglomerates depend on the strength of particle-particle forces when in surface contact. It is perhaps these surface forces which are the best known of the three phenomena.

Batel (ref. 83) has briefly discussed Van der Waals, capillary adhesion (water bridging), and Coulomb (electrostatic) forces between dust particles. He concludes that under "normal" conditions, capillary adhesion dominates. Štorch (ref. 84) has condensed industrial experience in the adhesion of dusts in his table 8, which lists various dusts in four classifications between non-adhesive and highly adhesive. His table emphasises the importance of moisture in adhesiveness, but also the chemical type and size of the dust. Schubert (ref. 85) has summarised calculated adhesive forces (of particles to a wall) in his figure 4, for a range of particle sizes, and has shown capillary adhesion at 5 times an average Van der Waals force, and 50 times a Coulombic force for an electrically conducting particle charged to 0.5 V. These forces were calculated for a 0.4 µm surface-surface approach. For larger particle-wall separations, capillary adhesion still dominated (for a 10 µm sphere) out to a 1 µm separation for high humidities, but dominated out to only 1 nm separation for 50 per cent relative humidity. Beyond these distances electrical forces predominated. The effect of surface roughness (asperities of the order 0.5 µm) appears overiding, and Coelho and Harnby (ref. 86) have emphasised the role adsorbed moisture layers (at lower humidities) and fully liquid bridges (at high humidities) play in covering this surface roughness so that bonding can occur.

It seems straightforward that moisture-held agglomerates are formed from small primary particles in a cyclone under normal humidities. These will be reduced in size on impaction with the wall, but will quickly form again in the shear layer (region of intense turbulence) if the concentration is high enough. Observations supporting this scenario are for example, Tengbergens; of large relative increases (2-3 fold) in collection efficiency of 0-3 µm dust fractions as the feed concentration was increased (ref. 2). Also Smigerski (ref. 87) found increasing efficiencies with *decreasing* particle size below 10 µm, at high humidities. He also found increasing efficiencies as the humidity was increased. For example, for 2 and 4 µm silica particles, the collection was increased from 20 to 80 per cent for a relative humidity increase from 15 to 90 per cent.

We can make two qualifications to this picture however. The first, a minor one, concerns the location of agglomerate break-up. The dispersion work (ref.56) discussed in 3.1 showed that double or triple-particle agglomerates were not broken significantly by gas interaction with the velocities that we have in most cyclones. However, we are concerned here with the breakup of perhaps 50-100 µm agglomerates to smaller agglomerates which will not approach the wall with appreciable velocity. Kousaka, Okuyama and Endo (ref. 88) have studied the breakup of such large agglomerates (composed of 0.1-1 µm primary particles) in an air flow. When their agglomerates rested on a glass plate, they found that the agglomerates cleaved internally, and explained the cleavage as due to bending forces induced by the high shear layer. Their results indicate (by extrapolation) that 50 per cent of 50 µm agglomerates should break off in the shear layer of a 20 m s^{-1} flow. If we ignore rotation, this corresponds to a 100 µm agglomerate entering such a shear layer. Thus some of the break-up may well occur before the larger agglo-merates reach the cyclone wall itself, but within perhaps $100\,\mu m$ from the wall.

The second qualification concerns electrical effects.

3.12   Electrical effects

Tengbergen (ref. 2) mentions the influence of dust electrical resistivity on collection. Dusts of high resistivity gave higher efficiencies than expected from his empirical calculations. Experiments with small sampling cyclones have shown more explicitly   electrical effects. Blachman and Lippmann (ref.89) collected very dilute aerosols in a D = 10 mm nylon cyclone, with average elementary charges of about 2000 per particle. They found significant increases in collection com-pared to that of a charge-neutralised aerosol, especially for smaller particles of 1-2 μm. On the other hand, Caplan, Doemeny and Sorenson (ref. 90) who used the same nylon cyclones and similar aluminium cyclones, found only that electrical charging of a coal dust increased and diminished the efficiency unpredictably, much beyond the normal scatter.

There can be little doubt that almost all dust is electrically charged by some means before entering a cyclone. For example dust from crushing operations has been shown by Kihlstedt (ref. 91) to have both positive and negative charges (up to ±100 elementary charges for 3 μm) with predominantly negative charges for particles less than 2 μm. Dust dispersed by an air blast has been found to be charged (both positively and negatively) on average 100 charges for 3 μm, and 10^4 charges for 10 μm dust (Mirgel (ref. 92) sugar, wood, soot, flour, silica). The convey-ing of dust along a duct causes charges predominantly of one sign, by contact and separation with the duct wall (e.g. 10^4 charges on 5 μm particles; Cole, Baum and Mobbs. ref. 93). Each of these examples appear to generate charges appro-ximately proportional to surface area for larger particles, probably concentrated on the surface asperities (Kottler, Krupp and Rabenhorst , ref. 94), and probably limited already by all larger surface charges having leaked away by local corona discharges before measurement was made. All the above examples are lower than the 100 charges per $(\mu m)^2$ surface assumed by Schuber (loc. cit) in his calculations.

The measurement of electrical charges of particles in cyclones has been done by Pearse and Pope (ref. 95) (who used a cyclone as a charging device) and Taubman and Popov (ref. 96). Both of these studies, done with large particles, gave average charges of the order of 10 per $(\mu m)^2$, which may be expected after an averaging of the higher individual positive and negative charges from air dis-persal etc.

We can now consider the influence of charging on collection. The most obvious mechanism for collection of dust via electrical charging is the migration towards the wall of charged dust particles by repulsion with other like-charged particles

in the cyclone volume (see Fig. 14). We can apply Gauss' theorem to a cylindrical
volume, relating the field at its surface to the charge Nq contained in it, and
if we assume Stokes viscous drag balances the electrical force, we can calculate
the radial velocity V_{Rq} of a particle at the surface of the cylinder. This has
been done by Foster (ref. 97) for example. We obtain

$$V_{Rq} = \frac{R \, q^2 \, N}{6\pi \, \varepsilon_o \, d_p \, \mu} \tag{73}$$

where q is the charge on each particle, N is the number of particles per unit
volume (assumed constant throughout the volume) and ε_o = 8.85 x 10^{-12} C^2 s^2
kg^{-1} m^{-3}. Applying this to the cylindrical portion of Hejma's cyclone with
R = 0.157 m, for his 1.5 μm dust, 3 g m^{-3} loading, N = 1.7 x 10^{12} m^{-3}, and
assuming 10 charges $(\mu m)^{-2}$ of particle surface, q = 70 charges, so that V_{Rq} = 7.4 mm
s^{-1}. This is too low to influence the collection significantly. Further down in
his cyclone, where the radius has reduced by half, but the concentration has
doubled, V_{Rq} is expected to remain the same.

A more favourable example may be a suspension of 5 μm clay particles at a
concentration of 20 g m^{-3} as collected in a D = 470 mm cyclone by Tengbergen.
If we use q = 10^4 charges with N = 10^{11} m^{-3}, V_{Rq} = 4.0 m s^{-1}, which is sufficient
to dominate the collection.

Soo (ref. 98) has also estimated the effect of charged particle repulsion in a
cyclone, and indicates that it can be the major collection mechanism for larger
cyclones.

Another mechanism is the short-range influence of the unlike charges on neighbouring particles. These unlike charges are expected to be larger than the like
charges considered above (methods of dispersal, and duct material, length, will
affect this). Those particles with unlike charges which are brought into close
promixity by the turbulent gas motion will have an enhanced collision rate (Fig.
4(b)). For example, two of the 1.5 μm particles considered above, with opposite
charges of 100, will approach each other with a velocity of 0.5 m s^{-1} when the
charges are separated by 0.1 μm. Only those charges actually on their near surfaces
will be important, whereas this calculation has assumed that all the charge is on
the near surface. For the 5 μm particles considered above with 10^4 charges, this
same velocity of 0.5 m s^{-1} is expected for a separation of 3.7 μm. In this case,
an appreciable electrical enhancement of collisions is expected (cf. an average
approach velocity of 2-3 m s^{-1}).

A somewhat less general possibility is the enhanced collision rate expected in
the high field at the dust rope. The layers of particles making up the rope and
on the heap are likely to have generated a locally intense field, as observed by
Maurer (ref. 99)(from 3 x 10^5 V m^{-1} up to the air break-down field 3 x 10^6 V m^{-1})

at the heap in a silo. These fields will oppose the outward radial travel of particles charged in like manner to those creating the field, and will enhance the radial motion of oppositely charged particles (Fig. 14(c)). If the small particles have the opposing charge, there is an enhancement of collisions. Freire and List (ref. 100) have investigated the collision rates for 2 and 10 μm particles with similar charges and electrical fields to those considered here. They found enhancement of collision rates by three orders of magnitude compared to non-electrical (gravity)collision. There is evidence from Kihlstedt (ref. 91) that smaller particles are predominantly negatively charged in crushing operations; how general this explanation can be is open to question.

Another mechanism affecting collection might be an enhancement of particle-particle bonding by Coulombic forces (Fig. 14(d)). Cross (ref. 100) has found that higher resistivity materials had considerably enhanced adhesion after electrical charging, even when the charge decayed to earth before the adhesion measurement was made. The amount of charge and the way it discharged played an important role. In view of the calculations done above, this may be the only important electrical effect for particles < 2 μm in a cyclone. Within a cyclone charging probably occurs in high-shear layers close to the wall or with the wall, and agglomerates formed in quieter turbulent regions (e.g. the bin) can be stronger as a result of the charging.

The method of electrical charging appears important, and cannot be overlooked. If the charges result from dust dispersal, or grinding operations, they will be of both signs, with possible collision and bonding enhancement. If the charges result from contacts made with a duct wall, they will be largely of the same sign, with particles avoiding collision, and the repulsion effect discussed above possibly important. If we have sufficiently strong like charges so that agglomeration is avoided, we can use eqn. 73 evaluated at r rather than R, and add the velocity V_{rq} in to the eqn. 66 to get a concentration profile across the cyclone where diffusion of particles in towards the centre balances the centrifugal and electrical repulsion fluxes. If the particles are small enough so that centrifugal forces can be ignored, and the concentration gradients are not large

$$\mathcal{D}_r \frac{d \ln c_j}{dr} = \frac{r \ q_j q N_r}{6\pi \ \varepsilon_0 \ d_{pj} \mu} + V_r \qquad\qquad (74)$$

where N_r is the total number concentration of particles, (averaged over r = 0 to r) and q is the average charge per particle. We have a new dimensionless group N_{qj},

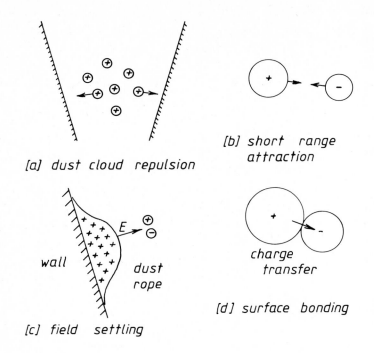

Fig. 14. Electrical effects on dust, promoting collection in a cyclone.

$$N_{qj} = \frac{q_j \, D \, q \, N}{d_{pj} \, \varepsilon_o \, V_i} \tag{75}$$

to characterise cyclone collection of small particles, where like-charging is expected. It can be noticed that we now have an efficiency which depends on dust *concentration*.

---

4. DUST-GAS INTERACTION AND PRESSURE DROP

Many investigators have shown that pressure drops across cyclones *decrease* with increased dust loading. This is in contrast with experience with pneumatic conveying in pipes. The pressure drop has been shown by Muschelknautz and Brunner (ref. 54) for example, to drop to a minimum at a dust concentration of about 1 kg m^{-3}, and rise again at higher concentrations. We will look closely at measurements done in the lower concentration range, to perhaps determine the mechanism of this dust-gas interaction, which also has a bearing on dust collection.

Tengbergen (ref. 2) has shown that for his cyclones, an increase from a low concentration to 50 g m^{-3} caused a ΔP reduction of about 20 per cent in his

high and low pressure-drop cyclones (models B and D in Fig. 1). In contrast, a
ΔP reduction of about 10 per cent was observed with an intermediate model. Yuu,
Jotaki, Tomita and Yoshida (ref. 102) observed a 35 per cent reduction in a short
cyclone (D = 296 mm) for concentration increases up to 50 g m^{-3}. The remarkable
aspect of their work was that only 0.2 g m^{-3} of dust was required in the feed to
reduce ΔP to the same value as at 50 g m^{-3}. Their work used dust with sizes from
.18 μm to 163 μm. They found a major drop (20 per cent) in the tangential velocity
V_θ within the cyclone with the introduction of 0.2 g m^{-3} dust, with smaller
further decreases in V_θ with higher dust loadings. They concluded that the ΔP
reduction was caused by the reduction in tangential velocity. Suspecting an
enhanced friction drag with the wall, they coated the wall (and the exhaust tube
outer wall) with a sticky liquid and then their dust. Some reduction in V_θ was
certainly observed, but it was not as large, and was of different distribution,
especially towards the centre. One pertinent aspect was that V_θ was reduced very
little at the outer wall, indicating that the circulation Γ or angular momentum
fed to the vortex sink was not materially altered, and wall effects could be
discounted.

Some further interesting observations have been made by Littlejohn and Smith
(ref. 103) in a small sampling cyclone. They recorded the flow at a given ΔP, and
found an increase of flow of about 25 per cent on introducing dust. The increased
flow was maintained after switching the dust feed off, and dropped to its "clean
air" level only after momentarily cutting off the air flow. If their cyclone bin
held more than 0.5g of dust, the air flow rate gradually rose again. Littlejohn
and Smith observed no such recovery with deeper bins, and related the flow behaviour
to a small amount of circulating dust they observed through transparent walls. This
recirculating dust remained after switching the dust feed off (in contrast to the
wall dust rope, which disappeared), was gone immediately after stopping the gas
flow, and gradually built up again on restoring the gas flow (presumably from
re-entrainment). They observed larger increases in air flow for larger particles
(in the range 50 to 200 μm) and for more angular particles. Their limit (0.25 g m^{-3})
above which the effect was observed, is close to that in the larger cyclone discussed
above.

The first effect one might consider to explain these observations is a "Reynold's
stress" due to the particles transporting angular momentum in a radial direction.
As can be seen from Fig. 8, gas towards the centre of a cyclone has a lower angular
momentum, and particles being transported out by centrifugal means will carry a
lower angular momentum with them, and slow the outer layers. This "particle
Reynold's stress" will add to the turbulent gas Reynold's stress, and enhance the
local turbulent viscosity. Most of the particles recirculating from the bin will
be carried up the central core and dispersed radially from there, providing a

roughly uniform concentration in the central region, as found by Reznik and Matsnev (Fig. 11). At high radial particle velocities $V_{rp} = \tau_p V_\theta^2 /r$, the Reynold's stress will be large; thus the effect will be largest close to the maximum V_θ for a given particle number concentration n_p. Towards the wall, V_{rp} drops, as also does $\partial V_\theta/\partial r$, reducing the expected Reynold's stress. This is in accord with the observations. We need to establish what recirculating mass concentration of particles is necessary for this effect.

The turbulent shear stress may be give as

$$\tau_{r\theta} = \nu_T \rho \; (\frac{\partial \overline{V}_\theta}{\partial r} - \frac{\overline{V}_\theta}{r}) \tag{76}$$

(cf eqn. 22) ignoring molecular viscosity. If we take a value of ν_T from eqns. 55, 70, for a cyclone, i.e.

$$\nu_T = 0.021 \; V_i D \tag{77}$$

then for $V_i = 9.8 \text{ m s}^{-1}$ in Yuu, Jotaki, Tomita and Yoshida's cyclone , and $r = 25$ mm, from their velocity profiles, $\tau_{r\theta} = 22 \text{ N m}^{-2}$.
The particle θ momentum flux in the r direction, $M_{\theta p}$ is given by

$$M_{\theta p} = n_p V_{rp} m_p r V_\theta \tag{78}$$

Using a particle mass concentration $C_p = n_p m_p$ and differentiating with respect to time, the rate of gain of particle momentum (equal to the rate of loss of gas momentum) per unit area, or the effective shear stress, $\tau_{\theta p}$ is given by

$$\tau_{\theta p} = C_p V_{rp} \; \frac{\partial}{\partial r} \; (V_{rp} r \; V_\theta) \tag{79}$$

It can be seen that this increases with the size of particle. Yuu, Jotaki, Tomita and Yoshida's PVC dust was of size 163 μm, with high enough radial velocity to ignore the correction for radial gas velocity, and again using their velocity profiles, $\tau_{\theta p} = \tau_{r\theta} = 22 \text{ N m}^{-2}$ when $C_p = 0.16 \text{ g m}^{-3}$. The observed 20 per cent drop in V_θ towards the centre is quite feasible with feed $C_p = 0.2 \text{ g m}^{-3}$ provided a similar concentration exists in the cyclone core. It should be noted that Reznik and Matsnev observed central C_p close to this figure where it seemed reasonable that much of the central dust was coming from the bin. Thus we appear to have a satisfactory explanation in terms of a particle Reynold's stress. The extent of ΔP reduction will depend on the suspended concentration C_p in the centre of the cyclone.

Parida and Chand (ref. 104) have numerically solved gas and particle momentum and continuity equations for a cyclone, assuming a constant eddy viscosity throughout the vortex (much lower than used here) and found a reduction of about 20 per cent in the maximum V_θ for a dust concentration of 0.2 g m^{-3}, supporting our simple analysis above.

As the dust concentration fed to the cyclone increases, the recirculated dust will also probably increase, and the radial drag of the larger dust will oppose the inward sink flow of gas. Crowe and Pratt (ref. 105) have modelled a uniflow cyclone loaded with 4-20 μm particles, using a finite-difference "tank and tube" model for the particle-gas suspension. For a feed concentration of 72 g m^{-3}, they found a 15 per cent reduction in the inward radial drift of gas. They point out that this is a mechanism for achieving a higher collection efficiency at higher dust concentrations.

---

5. CYCLONE RECOMMENDATIONS AND DESIGN CHANGES

In this chapter we have discussed the physical phenomena occurring largely in return-flow cyclones. On the basis of this discussion, it may be useful if some design and operating suggestions for return-flow cyclones are given here, with some assessment of recent cyclone-type collectors and modifications.

5.1 Return-flow cyclone recommendations

1. To achieve the maximum efficiency, the inlet design should be such as to avoid short circuiting of gas just under the exhaust lip. To avoid this, it appears that the inlet flow needs to stand off somewhat from the exhaust tube. Measurements of V_z across the radius above and below the exhaust lip and integration of flow, will give the best measure of this short circuiting for existing cyclones. An easy method for correction of this short circuiting may be the introduction of a small flow of clean air tangentially through the roof close to the exhaust. If it is given appreciable swirl, it will correct both for the roof boundary layer leakage, and the slowing down of gas on the outer surface of the exhaust. (This idea has already been used with cyclone combustion chambers; see Ross (ref. 106) and Gyarmathy (ref. 107)).

2. A long cyclone achieves higher efficiency by removing the recirculating dust (bin-cone exchange) from the exhaust mouth. The successful prediction of concentration profiles across the cyclone (except for the higher recirculation contribution at the centre) means that dust has radially equilibrated for each axial position, with turbulent diffusion dominant. To achieve lower recirculating dust concentrations, and higher efficiencies, large deep air-tight bins are recommended.

A number of workers have used an inverted cone situated centrally in the mouth of the bin to discharge the dust around the annulus, and prevent re-entrainment.

Ter Linden described one (ref. 18), Barth and Leineweber (ref. 108) have given many examples from industry where they have been used, and Krambrock (ref. 109) give some optimised dimensions (see Fig. 15(a). Krambrock shows the necessity of an enlarged dust exit diameter (at least equal to the exhaust diameter) with a correspondingly large cone to avoid the vortex end attaching itself to the wall, with consequent very low efficiencies. It is expected that some stabilising of the vortex on the tip of the cone may occur. Krambrock claims that improvements in efficiency of 10 to 20 per cent may be made, and this may well be so for particles > 10 μm, but for small particles where agglomeration-electrical charging is important, the necessary recirculation is not provided, and efficiency may fall.

One design, successful with dust < 10 μm, incorporates an inverted cone within the conical section (see Fig. 15(b)) with a hole in the centre of the inverted cone (see Ryabchikov, Amerik and Karpukhovich ref. 110). This hole allows air through to feed the vortex and stabilise it, and probably confines the re-circulation of fine dust to the portion below the inverted cone. They reported a 93.2 per cent efficiency for a dust with 33 per cent mass < 10 μm, for D = 1.4 m, and with a 180° wrap-around entry, achieved this for a ΔP a little over half that of other cyclones with the same efficiency. The design is useful also in that the pressure in the bin is much higher; the effect of leaks is expected not to be so drastic, and it is suitable for use e.g. with a stand-pipe return to a fluid bed.

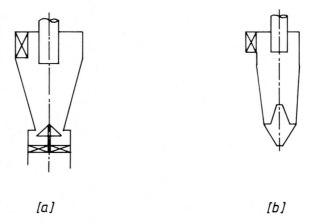

[a]   [b]

Fig. 15 (a) Optimal dimensions of the Bayer AG cyclone (b) ST$_s$ KN cyclone

3. The vortex break-down gas bubble identified in this chapter as lying just within the exhaust tube appears to affect both the efficiency of a cyclone and its pressure drop. The instabilities in its radial position will probably dictate the level of radial turbulence, and hence the radial concentration of dust. It appears to be worthwhile (see 2.4.3) to insert a machined liner into the exhaust, heavy enough to keep the exhaust inner surface circular to within an 0.5 mm eccentricity and perhaps extending from the exhaust lip up one diameter, sufficient to stabilise the gas bubble.

Also some attempt at removing the dust collected at the exhaust wall opposite the gas bubble, may be beneficial. This dust can collect here because just within the exhaust, the swirl is still high, but the radial velocity is *outward* (to avoid the bubble) or zero. Stairmand shows (ref. 1) a withdrawal gap in the exhaust, some diameters along. Also a British manufacturer (ref. 111) appears to find withdrawal about 1 D_e up from the exhaust lip and reinjection into the feed worthwhile. Fig. 16 shows their double annulus arrangement, with which they claim 80 per cent recovery of $5 \mu m$ fly ash (cf. 60 per cent with a normal " high efficiency" cyclone operating with half the pressure drop.)

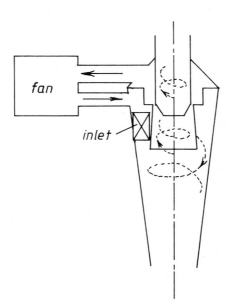

Fig. 16 Withdrawal and reinjection.

A simpler solution may be a countercurrent swirling clean air flow introduced through an annular gap opposite the gas bubble (see Fig. 17). This is intended to reinject the dust back into the cyclone with sufficient V_θ that it will join the downward moving material. Also it may prevent lip leakage. Trials done by this author with the insert shown in Fig. 17, made with the one geometry shown there, have given a promising result. Collection of the alumina dust detailed in the appendix gave mean efficiencies of 99.03 (unmodified) and 99.34 per cent (with insert) for collection in the cyclone of Abrahamson, Martin and Wong (ref. 43). This improvement represented about 30 per cent reduction in emission, and although based as yet on a limited number of runs, is statistically significant to the 0.05 level. The pressure drop was negligibly different with the insert, the inlet velocity was reduced by 8 per cent, and the exhaust flow increased by 35 per cent (of the inlet air) being brought in through the insert. No attempt was made to optimise the configuration, but it appears worthy of further investigation.

Some success has been achieved recently in attempts to recover some of the rotational energy lost in the exhaust, and so reduce the cyclone ΔP. Browne and Strauss (ref. 112) appear to have replaced (inadvertently) the gas bubble with a torpedo-shaped constrictor-diffuser, and with suitable vanes have reduced the ΔP by 22 per cent, with no drop in collection efficiency (Olszewski and Strauss ref. 113). Pervov (ref.114) has devised a similar drop-shaped

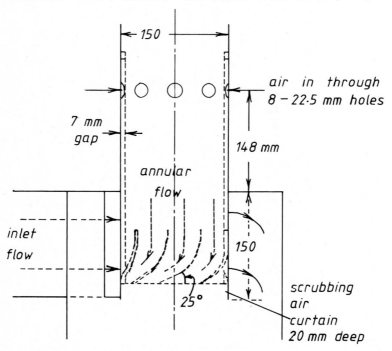

Fig. 17   Dust reflux insert.

"helically bladed unswirling unit" which he has tested with dust of mean size 9 μm. In contrast to other devices tested, his helically bladed device held the same efficiency or improved it by 2-3 per cent, while reducing ΔP by 16-24 per cent. It appears suitable for installing on single and multiple cyclones, and also reduces the vacuum in the hopper. It is expected that the vortex is stabilised on the forward point of the drop-shaped centre-body.

4. More attention needs to be given to the conditioning of dust before it reaches a cyclone, especially for dust < 10 μm in size. Relative humidities >30 to 50 per cent may improve collection significantly, as also may gaseous conditioners used with electrostatic precipitators (e.g. amines) apparently to improve agglomeration. The difference in electrical charging from air jet dispersal and contact with ducts may give some differences between laboratory trials and plant experience. It may be **worthwhile** experimenting with different surfaces in the approach dust, and even with alternating surfaces with different work functions within the cyclone, to achieve high concentrations of oppositely charged particles. The possibility of adding small amounts of an easily collected "sticky" dust to a more difficult dust, should sometimes be investigated.

5. For the engineer wishing to choose cyclones for a given dust collection job, he cannot do better than to consult Štorch (ref. 84) who details factors beyond the efficiency and pressure drop considered here, which determine the *reliability* of **operation. These are -** mechanical defects, clogging and wear.

5.2 Uniflow high-efficiency cyclones and concentrators

We have seen that there is some advantage in recirculating fine dust prior to collection if the recirculation can be removed far enough from the gas exhaust. **Recently a high efficiency of** *concentration* **was obtained in a through-flow cyclone, with the concentrated flow collected by a much** smaller cyclone operating at higher concentration. Fig. 18 shows such a cyclone (Kalmykov ref. 115) with the dusty gas entering through swirling vanes at the bottom, travelling upwards around a central cone, and being split between a small annular flow (8-15 per cent) with the dust, and a cleaned central flow. The level of turbulence is likely to be reduced compared to that in a return-flow cyclone, and the gas bubble will be stabilised by the point of the cone. Ignat'ev and Kalmykov (ref. 116) have investigated a 360 mm dia. example, and found excellent collection of the concentrator-collector combination ($\sqrt{Stk_{50}}$ = 0.063, $\sqrt{Stk_{80}}$ = 0.22, less than Stairmand's cyclone A, Fig. 1), with low pressure drops. **An** advantage of this unit appears to be its compactness.

Another **novel** device is the rotary-flow dust collector, which **imparts a**

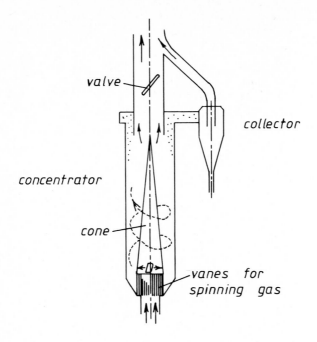

Fig. 18 Cyclone concentrator collector combination. Stk based on D

rotational flow to the dust gas to be cleaned, and then increases the angular
velocity by bringing a secondary clean air flow in at high velocity through
tangential inlets at the walls (see Fig. 19). The extra complications and
ranges of variables are probably the reason why some workers have not had success
with this design (e.g. Ciliberti and Lancaster, ref. 117). However Hikichi and
Ogawa have thoroughly evaluated its potential and find for a 150 mm dia. unit
with L/D = 3 that d_{p50} of 1.0 µm is easily obtained, and values as low as 0.4 µm
were obtained, with d_{p95} = 1.5 µm. However Hikichi and Ogawa's unit is large
for a given throughput. The high efficiencies probably result from a stable
vortex, with higher Γ at large r, and from some reverse flow at the wall, towards
the dust bin.

5.3 <u>Rotating-wall collectors</u>

   We have seen in 2.2 that lower Γ or angular momentum at larger radius will
cause instability and greater momentum (and dust) exchange between layers in a
vortex. Another method to avoid these radial instabilities is to rotate the
outer wall *faster* than the gas. This procedure should also remove Görtler
vortices (2.4.5) and their attendant "dust ropes", with the wear problems sometimes
encountered (Štorch ref. 84). However, it will not eliminate a precessing vortex
core, and other precautions already mentioned may help here.

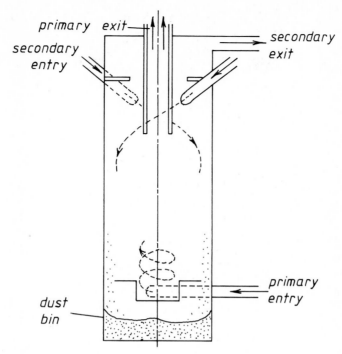

Fig. 19 Rotary flow separator

Razgaitis, Guenther and Bigler (ref. 119) have begun work in this direction by rotating a cylindrical section within a cyclone. Their rotation for most of the work did not exceed that of the gas, but they measured a drop of turbulent intensity V'_θ/V_θ from 15 per cent down to 5 per cent. On the other hand, the Reynold's stress ratio $\overline{V'_r\ V'_\theta}/V^2_{\theta\ max}$, did not show a reduction. Their pressure drop increased with rotation, as expected with a *lower* v_T in the radial direction. In practice, a large gain in dust collection efficiency achieved by rotation was possibly almost lost by the mismatch between their cylinder and the cyclone wall. They reported a gain of collection efficiency from 85 to 89 per cent for a wall velocity equal to that of the gas, and to 94 per cent for a higher velocity.

Schmidt (ref. 120) has compared two designs of rotating-element cyclone with normal stationary cyclones, using the plot of pressure-drop number ζ (eqn. 2) versus B, used by German workers to characterise cyclones (see Strauss ref. 49). B is essentially a ratio of the relaxation time τ_p of the particle calculated to be (just) collected, to a time of passage of gas through the cyclone. On this basis, Schmidt showed that his rotating-element cyclones collected finer particles for a given pressure drop than a conventional cyclone. His results are interesting because he rotated the *exhaust tube* in the more successful model. In line with the work mentioned in 5.1 on further collection in the exhaust, the use of a rotating back-flushed exhaust may have potential for high efficiency collection.

Kanagawa, Takahashi and Yokochi (ref. 122) have rotated both outer and inner walls of an annular cyclone collector, which also acted as the gas blower. This concept appears to have much promise even though blowing efficiencies were low. Values for d_{p50} less than 1 μm were relatively easily obtained with the device shown in Fig. 20.

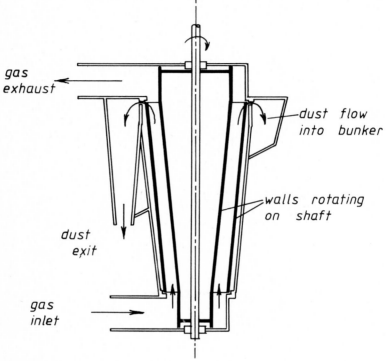

Fig. 20   Double rotor collector

---

6. CLOSING REMARKS

This author did not realise the number of aspects of cyclone collection which were amenable to a quantitative analysis (although only one at a time!), before writing this chapter. The cyclone has been almost passed by as a single dust collector in the face of more stringent emission levels, but we have seen in this chapter the promise of a large improvement in efficiency over traditional cyclone designs. It is hoped that the mechanisms discussed here will give readers a firmer basis on which to assess new (and old) designs, and will encourage them to devise and test some new variants themselves.

ACKNOWLEDGEMENTS

I thank J. W. Redmayne for his rapid and efficient library service, T. Berry for help in photography, G. Grey and W. Earl for great labours in developing the multiflash unit, and M. Farrier, N.Z. Aluminium Smelters Ltd., for the alumina dust and analysis.

Mrs B. Singh has worked hard to bring an untidy and labyrinthine script into this form, and I wish to record my gratitude.

SYMBOLS

Latin

a   empirical index for Re

A   constant in Burger's vortex, s^{-1};

  or area of cross-section, m^2

C   radial turbulent Re;

  or dust concentration, kg dust m^{-3} gas

d_p particle diameter, m

D   cyclone diameter, m*

h   exhaust to hopper height, m

  or as in Fig. 7

L   length, m

m   swirl velocity index

M   momentum flux, N m^{-2}

M_p mass of particle, kg

n   number of particles

N   number concentration, m^{-3}

p   pressure, N m^{-2}

q   electric charge, C

Q   gas flow rate, $m^3\ s^{-1}$

r   radius, m

R   cyclone radius,   m*

Re   Reynolds number

Stk cyclone – particle Stokes number

t   time, s

T   temperature, K

  or jet time, s'

V   velocity, m s^{-1}

z   axial distance, m

Greek

Γ   gas circulation, $m^2\ s^{-1}$

Δ   difference

ε turbulent power dissipation, W kg^{-1}
 or electrical permittivity, C^2 s^2 kg^{-1} m^{-3}

ζ pressure drop number

η mass efficiency of collection, collected/fed

θ tangential component

Λ viscosity constant

μ gas viscosity, Pa s

ν kinematic gas viscosity, m^2 s^{-1}

ξ dimensionless radius

ρ gas density kg m^{-3}

τ shear stress, N m^{-2}
 or eddy relaxation time, s

ψ particle-particle Stokes number

Ω swirl number

Subscripts

b location

e exit

i inlet

j of the jth size

o outlet, or surroundings
 or outer scale

p particle

p50 particle captured 50 per cent

r radial component

rel relative

Rq at radius R due to charge q

T turbulent

z axial component or direction

Superscript

' fluctuation

---

* If not subscripted, refers to cyclone barrel.

APPENDIX I

FLASH LAMP SPECIFICATIONS

The flash lamp was built up from two flash bulbs for Strobotac Stroboscopes, made by Gen Rad, Concord, Mass. The bulbs were filled with Xenon at 0.5 bar. The flash bulbs were separately supplied from 900 V d.c. stored in 0.1 μF capacitors (40 mJ available), separately triggered with a variable delay of 5, or 10 μs. The light pulse width at 1/3 peak height was 1.4 μs.

Light from each bulb was collected by 70 mm dia. front=silvered parabolic mirrors, and either passed directly through, or reflected from, a half silvered mirror, so that a double flash was obtained with the same optical path into the camera.

APPENDIX II

ALUMINA DUST USED FOR PHOTOGRAPHY AND EFFICIENCY TESTS

Multicyclone hopper dust

1. Particle size

Particle Size Range (μm)	% Wt Fraction in range
0-2	0.8
2-4	4.0
4-6	7.4
6-10	18.8
10-15	19.0
15-20	13.0
20-30	16.0
30-40	8.0
40-70	10.5
> 70	2.5

2. Specific Gravity

Size range (μm)	Result
0-6	2.9
6-18	2.8
18-40	3.2
> 40	3.0

3. Fluoride

8.0% w/w

REFERENCES

1 C.J. Stairmand, Trans. Instn. Chem. Engrs., 29(1951)356-372.
2 H.J. van Ebbenhorst Tengbergen, De Ingenieur, 77 No.2(1965) Wl-W8.
3 A.V. Kalmykov, Yu. M. Afanas'ev and N.G. Shipunov, Thermal Engng., 22 (1975-76) 70-73.
4 W. Barth, Brennstoff-Wärme-Kraft, 8 No.1(1956)1-9, as discussed by W. Batel, Staub-Reinhalt. Luft, 32 No.9(1972) 1-7(Engl. transl.)
5 D. Leith and W. Licht, in Air Pollution and its Control AIChE Symposium Series 1972, No.126, pp.196-206.
6 J.O. Parent, Trans. Am. Inst. Chem. Engrs., 42(1946) 989-999.
7 L.C. Whiton, Chem. & Metall. Engrg., 39(1932) 150-151.
8 W.B. Smith, R.R. Wilson and D.B. Harris, Environ. Sci. & Technol. 13(1979) 1387-1392
9 T.M. Knowlton and D.M. Bachovchin, in Coal Process Technol. 4, Editors of Chem. Eng. Prog. (1978), pp.122-127.
10 A.C. Stern, K.J. Caplan and P.D. Bush, Cyclone Dust Collectors, Am. Petrol. Inst.,Washington, D.C. (1955), p.39.
11 B. Kalen and F.A. Zenz, in R.W. Coughlin et al.(eds.) Recent Advances in Air Pollution Control, A.I.Ch.E Sympos. Ser. No.137, Vol.70, 1974, pp.388-396.
12 T. Chan amd M. Lippmann, Environ. Sci. & Technol. 11(1977)377-382.
13 J.M. Beeckmans and C.J. Kim, Canad. J. Chem. Eng., 55(1977)640-643.
14 N.I.Zverev and S.G. Ushakov, Thermal Engng, 16(4)(1969)81-84.
15 V.A. Reznik, N.N. Prokofichev and V.V. Matsnev, Thermal Engng., 18(5)(1971)68-73.
16 V.V. Matsnev and S.G. Ushakov, Thermal Engng. 23(9)(1976)80-82.
17 C. Doerschlag and G. Miczek, Chem. Engng., 84,(4)(1977)64-72.
18 A.J. Ter Linden, Proc. Inst. Mech. Engrs.(Lond.) 160(1949)233-251.
19 N. Cherrett, J. Inst. Fuel, 35(1962)245-250.
20 C.M. Peterson and K.T. Whitby, ASHRAE J., 7(5)(1965)42-49.
21 W. Rausch, Verfahrenstechnik,3(5)(1969)214-221.
22 R.V. Avant, C.B. Parnell and J.W. Sorenson, Paper no. 76-3543, presented at the 1976 Winter Meeting Am. Soc. Ag. Engrs., Chicago, Dec. 1976.
23 H.E. Ayer and J.M. Hochstrasser, Paper in Aerosol Measurement Workshop Proceedings, Gainesville, Fla., March 1976, pp.70-79.
24 J.M. Hochstrasser, Doctoral Dissertation, Univ. of Cincinnati 1976 (Dept. Environ. Health).
25 A.K. Garg and S. Leibovich Phys. Fluids, 22(1979) 2053-2064..
26 J.M. Burgers Advances in Appl. Mech., 1(1948) 198.
27 M.S. Uberoi J. Fluid Mech., 90(1979) 241-255.
28 J.K. Harvey J. Fluid Mech., 14(1962)585-592.
29 J.H. Faler and S. Leibovich Physics Fluids 20(1977) 1385-1400.
30 Lord Rayleigh Proc. Roy. Soc;, A93(1916) 148-154.
31 M.G. Hall, The Structure of Concentrated Vortex Cores, in Progress in Aeronaut. Sci. 7(1966)53-110.
32 S.Ito, K. Ogawa and C. Kuroda, J. Chem. Eng. Japan, 13(1980)6-10.
33 J.H. Faler and S. Leibovich J. Fluid Mech., 86(1978)313-335.
34 N. Syred and J.M. Beér Comb. Flame, 23(1974)143-201.
35 D.G. Lilley AIAAJ, 15(1977)1063-1078.
36 C.R. Church, J.T. Snow, G.L. Baker and E.M. Agee, J. Atmos. Sci., 36(1979) 1755-1776.
37 G.L. Baker and C.R. Church, J. Atmos. Sci., 36(1979) 2413-2424.
38 J.L. Smith Trans. ASME., J. Basic Eng., 84(1962)602-608.
39 A.A. Pervov Chemical & Petroleum Engng., 10(1974)898-900. (Transl. of Khim. Neft. Mashin.)
40 J. Hejma Staub-Reinhalt 31(1971)22-28.
41 R.M. Alexander Proc. Australian Min. & Metall. N.S. Nos.152-153 (1949)203-228.
42 R. Razgaitis, D.A. Guenther and M.J. Bigler Paper 79-WA/APC-8 presented at Winter Ann. Meeting N.Y. Dec. 1979 of Am. Soc. Mech. Eng.
43 J. Abrahamson, C.G. Martin and K.K. Wong, Trans. Inst. Chem. Engrs. 56(1978) 168-177.

44 V.A. Reznik and V.V. Matsnev, Thermal Engng., 18 No.12(1971)34-39.

45 J. L. Smith, Trans. ASME, J. Basic Eng., 84(1962)609-618.

46 W.R. Schowalter and H.F. Johnstone A.I.Ch.E.J., 6(1960)648-655.

47 F. Löffler and P. Meissner in P. Davalloo et al. (Eds) Proc. 1st Iran. Congress Chem. Engng., Elsevier, Amsterdam, 1973, pp.405-413.

48 C.J. Stairmand Engineering, 168(1949)409-412.

49 W. Strauss, Industrial Gas Cleaning, 2nd Edn., Pergamon, Oxford, 1975, p.227, 243,249.

50 S.A. Tager, Thermal Engng, 18 No.7(1971)120-125.

51 E.D. Baluev and Yu.V. Troyankin Thermal Engng. 14 No.2(1967) 99-105.

52 C.B. Shepherd and C.E. Lapple Ind. & Eng. Chem., 31(1939) 972-984(Run 5, b=3, Fcv).

53 M.I.G. Bloor and D.B. Ingham Trans. Inst. Chem. Engrs., 53(1975)7-11.

54 E. Muschelknautz and K. Brunner, Chem. Ing. Tech., 39(1967)531-538.

55 D. Coles, J. Fluid Mech., 21(1965)385-425.

56 Y.Kousaka, K. Okuyama, A. Shimizu and T. Yoshida, J. Chem. Engng. Japan, 12(1979)152-159.

57 T. Yoshida, Y. Kousaka and K. Okuyama, Aerosol Science for Engineers, Power Co. Ltd., Tokyo, 1979, p.175.

58 G. Jimbo and S. Fujita, Paper in Proc. of Powtech '71: Int. Powder Technol. & Bulk Gran. Solids Conf. A.S. Goldberg (ed.) Powder Advisory Centre, London, 1971, p.155-161.

59 M.A. Delichatsios and R.F. Probstein, J. Colloid & Interface Sci., 51(1975) 394-405.

60 J. Abrahamson Chem. Eng. Sci. 30(1975)1371-1379.

61 K. Okuyama, Y. Kousaka and T. Yoshida J. Aerosol Sci., 9(1978)399-410.

62 J.J.E. Williams and R.I. Crane J. Mech. Eng. Sci. 21(1979) 357-360

63 S.L. Soo, Powder Technol., 7(1973)267-269.

64 B.Dahneke, J. Coll. Interface Sci. 51(1975)58-65.

65 J.W. Cleaver and B. Yates, Chem. Eng. Sci. 30(1975)983-992.

66 M.S. El-Shobokshy and I.A. Ismail, Atmos. Environ. 14(1980)297-304.

67. J.W. Cleaver and B. Yates. J. Coll. Interface Sci., 44(1973)464-474.

68 P.A. Arundel, S.D. Bibb and R.G. Boothroyd Powder Technol., 4(1970/71)302-312.

69 R.G. Boothroyd and P.J. Walton, Ind. Eng. Chem. Fundamentals 12(1973)75-82.

70 R. Zisselmar and O. Molerus, Chem.Eng. J., 18(1979)233-239.

71 V.E. Maslov, V.D. Lebedev, N.I. Zverev and S.G. Ushakov Thermal Engng., 17(4) (1970)133-136.

72 T. Mizushina and F. Ogino J. Chem. Eng. Japan, 3(1970) 166-170.

73 R.B. Bird, W.E. Stewart and E.N. Lightfoot "Transport Phenomena" Wiley, New York, 1960, pp.379,629.

74 R.G. Boothroyd Trans. Instn. Chem. Engrs., 45(1967)297-310.

75 J. Weinstock Phys. Fluids, 21(1978)887-890.

76 R.S. Brodkey in "Turbulence in Mixing Operations" Academic, New York, 1975, pp.80, 84.

77 S. Yuu, N. Yasukouchi, Y. Hirosawa and T. Jotaki A.I.Ch.E.J., 24(1978)509-519.

78 I. Tani J. Geophysical Res., 67(1962)3075-3080.

79 E. Muschelknautz Chem. Ing. Tech., 44(1972)63-71.

80 K. Adachi, S. Kiriyama and N. Yoshioka Chem. Eng. Sci., 33(1978)115-121.

81 J. Yerushalmi, D.H. Turner and A.M. Squires Ind. Eng. Chem., Process Des. Dev. 15(1976) 47-53.

82 C.D. Hill and A. Bedford Phys. Fluids 22(1979)1252-1254.

83 W. Batel, Dust Extraction Technology, Engl. transl. by R. Hardbottle, Technicopy, Stonehouse, Glos. England, 1976 p.19.

84 O. Štorch et al. Industrial Separators for Gas Cleaning, Elsevier, Amsterdam, 1979, p.45, 120.

85 H. Schubert Chem. Ing. Tech., 51(1979)266-277.

86 M.C. Coelho and N. Harnby Powder Technol., 20(1978)201-205.

87 H.J. Smigerski "Die Feinstaubagglomeration in Fliehkraftenstaubern", Fortschrittberichte der VDI Zeitschriften, 3 No.30, August 1970, VDI-Verlag, GMBH, Düsseldorf.

88 Y. Kousaka, K. Okuyama and Y. Endo, J. Chem. Eng. Japan, 13(1980)143-147.

89 M.V.Blachman and M. Lippmann Am. Ind. Hygiene Assoc. J., 35(1974)311-326.
90 K.J. Caplan, L.J. Doemeny and S.D. Sorenson, Am. Ind. Hygiene Assoc. J., 38(1977)162-173.
91 P.G. Kihlstedt "Distribution of Electrical Charges in Mineral Dust" Dechema Monographien 79 No.1589-1615, pt B(1975?)95-114.
92 K.H. Mirgel, VDI Berichte 19(1957)49-57.
93 B.N.Cole, M.R. Baum and F.R. Mobbs Proc. Inst. Mech. Engrs., London, 184 Part 3C(1969-70)77-83.
94 W. Kottler, H. Krupp and H. Rabenhorst Zeits. Angewandte Physik, 24(1968) 219-223.
95 M.J. Pearse and M.I. Pope Powder Technol., 14(1976)7-15.
96 I.S. Taubman and B.G. Popov Soviet Plastics No.10(1971)77-79.
97 W.W. Foster Brit.J. Appl, Phys. 10(1959)206-213.
98 S.L. Soo "Some Basic Aspects of Cyclone Separators" in Proc. 1st Int. Conf. Particle Technol., 1973, IIT Research Inst., Chicago, pp.10-16.
99 B. Maurer Ger. Chem. Eng., 4(1979)189-195.
100 E. Freire and R. List J. Meteor. Sci., 36(1979)1777-1787.
101 J.A. Cross "Electrostatic Effects in the Adhesion of Powder Layers" in Surface Contamination-Genesis, Detection + Control Ed. K. Mittal, Vol.1, Plenum, 1979, pp.89-102.
102 S. Yuu, T. Jotaki, Y. Tomita and K. Yoshida, Chem. Eng. Sci., 33(1978)1573-1580.
103 R.F. Littlejohn and R. Smith Proc. Instn. Mech. Engrs., 192(1978)243-250.
104 A. Parida and P. Chand, Chem.Eng. Sci., 35(1980) 949-954.
105 C.T. Crowe and D.T. Pratt Computers and Fluids 2(1974)249-260.
106 D.H. Ross "An Experimental Study of Secondary Flow in Jet Driven Vortex Chambers" Report No. ATN-64(9227)-1 Aerospace Corp., Unclass. rep. No.AD 433052, Clearinghouse Fed. Sci. Tech. Info. U.S. Dept. Commerce.
107 G. Gyarmathy "Optical Density Measurements in a High Speed Confined Gaseous Vortex" Pap. No.68-694, A.I.A.A. Fluid and Plasma Dynamics Conf., Los Angeles, 1968.
108 W. Barth and L. Leineweber Staub 24(1964)41-55.
109 W. Krambrock Chem. Ing. Tech. 51(1979)493-496.
110 S.Y. Ryabchikov, B.K. Amerik, and D.T. Karpukhovich Chemistry & Technology of Fuels and Oils 10 No. 10 (1974) 781-785. (Transl. from Khimiya i Tekh. Topliv i Masel).
111 Collectron Fine Particle Recovery Units, Collectron Ltd., 5 Greenhills Rd., Charlton Kings, Cheltenham, Glos., U.K.
112 J.M. Browne and W. Strauss Atmospheric Environ. 12(1978)1213-1221.
113 A. Olszewski and W. Strauss Atmospheric Environ.12(1978)1559.
114 A.A. Pervov Chemical and Petroleum Engng., No.1(1975)24-27 (Transl. from Khimicheskoe i Neft. Mash.)
115 A.V. Kalmykov Thermal Engng., 17(4)(1970)91-95.
116 V.I. Ignat'ev and A.V. Kalmykov Thermal Engng., 13(6)(1966)49-53.
117 D.F. Ciliberti and B.W. Lancaster, Chem. Eng. Sci., 31(1976)499-503.
118 T. Hikichi and A. Ogawa Bulletin Jap. Soc. Mech. Engrs . 22(1979)815-824.
119 R. Razgaitis, D.A. Guenther and M.L. Bilger "Turbulence Suppression in a Cyclone Separator by Means of a Rotating Insert" pap. 79-WA/APC-8 presented at Winter Ann. Meet. Am. Soc. Mech. Engrs., New York, Dec.2-7, 1979.
120 P. Schmidt Chem. Ing. Tech. 47(1975)107.
121 A. Kanagawa, T. Takahashi and A. Yokochi, Int. Chem. Engng., 18(1978)627-633.

FILTRATION OF GASES IN FLUIDISED BEDS

ROLAND CLIFT, MOJTABA GHADIRI and KAILAI V.THAMBIMUTHU
Department of Chemical Engineering, University of Cambridge,
Pembroke Street, Cambridge, CB2 3RA, England.

I. INTRODUCTION

The idea of using a fluidised bed to remove dust or mist particles from a gas appears to have originated with Meissner and Mickley (ref.1). The basic concept, shown schematically in Fig.1, involves using the gas to be filtered to fluidise a bed of a suitable granular collection medium. The mobility of the bed material enables continuous removal and replacement of the collector, and interest in fluidised filtration first arose from this possibility of continuous operation. More recent research has shown that a fluidised bed can give high filtration efficiencies for gas-borne particles a few microns in diameter. Current interest in the technique therefore centres on its possible use for tertiary gas cleaning, especially under difficult process conditions.

The following review concentrates on the fundamental processes occurring in fluidised filters. Section II gives a brief summary of filtration in fixed granular beds, to define terms and concepts used in analysis of fluidised filtration. Attention is restricted to "stationary" filtration, in which the collection medium is the granular material itself, without discussing "non-stationary" effects caused by structural changes resulting from dust accumulation within the filter. Section III sets out an analysis of filtration in a fluidised bed, derived from conventional models of filtration in fixed beds and reaction in fluidised beds. The interpretation of available experimental results in terms of this model is summarised in Section IV. Section V reviews two variations which have been proposed on the simple configuration of Fig.1: electrofluidised filters, and rotating fluidised beds. Actual and potential applications of the technique are discussed very briefly in Section VI.

In a filter such as a fluidised bed where the collector is mobile, it is useful to distinguish between two distinct processes. "Collection" refers to the process whereby a gas-borne particle is brought into contact with a collector granule.

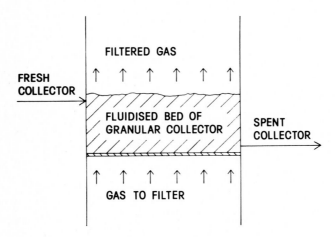

Fig.1 Fluidised bed filter (schematic)

"Retention" refers to phenomena which retain the particulate on the collector without being re-entrained. For effective filtration, the efficiencies both of collection and retention must be reasonably high. Most laboratory studies have ensured essentially complete retention either by using a mist as the challenging aerosol or by precoating the collector with a thin layer of a non-volatile liquid. The analysis in Sections II-IV refers to the case where all particles are retained, so that filter performance is determined by collection. Retention is a practical problem, discussed in Section VI.

II. FILTRATION IN FIXED GRANULAR BEDS

II.1 Mechanisms of filtration

It is conventional (ref.2) to distinguish between four mechanical processes causing filtration in a fixed granular bed or a fibrous filter. These processes are illustrated schematically by Fig.2. In each case, a gas-borne aerosol particle, of diameter d_a, is assumed to be filtered if it contacts the surface of the collector particle, of diameter d_c.

Diffusional collection results from transport of aerosol to the surface of the collector by Brownian motion through the gas. For the applications of interest here, in which aerosol around 1μm is filtered from gases at ambient and higher pressures, the mean free path of gas molecules is typically comparable to or smaller than the aerosol diameter. Under these circumstances, the Brownian diffusivity, D, can be estimated from the Stokes-Einstein equation (see ref.2):

$$D = F k_B T/3\pi\mu d_a \qquad (1)$$

For particles which are much smaller than the mean free path but large by comparison with gas molecules, a result obtained by Langmuir should be used:

$$D = \frac{4k_B T}{3\pi P d_a^2} \left[\frac{8RT}{\pi M}\right]^{\frac{1}{2}} \qquad (2)$$

In eqs. (1) and (2), k_B is Boltzmann's constant (i.e. $1.380622 \times 10^{-23} JK^{-1}$), T and P are the gas temperature and pressure, μ and M are the gas viscosity and molecular

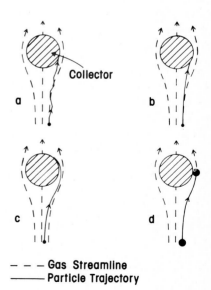

Fig.2. Schematic representation of collection mechanisms in fixed beds:
(a) Diffusion (b) Inertia (c) Gravity
(d) Interception

weight, and R is the universal gas constant. The dimensionless coefficient F in eq.(1) is the Stokes-Cunningham "slip correction factor", which allows for the fact that the viscous drag experienced by a particle whose dimensions are comparable with the molecular mean free path is less than that on a particle in a fluid which behaves as a continuum (see ref.3). For spherical particles, the slip correction factor can be estimated (ref.4) as

$$F = 1 + (\lambda/d_a)\left[2.514 + 0.8\exp(-0.55d_a/\lambda)\right] \tag{3}$$

where λ is the molecular mean free path. Modifications to eqs.(1) and (3) for non-spherical particles are considered in refs.(3) and (5). For air, λ can conveniently be estimated from a correlation due to Beard (ref.6):

$$\lambda = 215\,\mu\,T^{\frac{1}{2}}/P \tag{4}$$

where λ is in μm, T in K, and P in bars (i.e. units of 100 kP). For other gases, λ can be calculated from the general result (see ref.6):

$$\lambda = \frac{\mu}{0.998}\sqrt{\frac{\pi}{2\rho_g P}} \tag{5}$$

where ρ_g is the gas density.

Typical values for the slip correction factor, F, and Brownian diffusivity, D, are shown in Fig.3. Two points are particularly worthy of note. For the range of aerosol size of interest, Brownian diffusivity is several orders of magnitude smaller than typical molecular diffusivities in gases. The effect of temperature on D is strong, primarily through T and μ in eqn.(1), but pressure only affects F. Therefore the influence of pressure on D is much weaker, and is only perceptible for temperatures well above ambient and aerosol particles smaller than typically 2μm (ref.7).

Inertial deposition occurs when the relatively large inertia of an aerosol particle causes its trajectory to deviate from the gas streamlines and hence contact the surface of the collector (see Fig.2b). Conventionally (ref.2), inertial effects are correlated in terms of the Stokes number:

$$St = F\rho_a d_a^2 U/9\,\mu\,d_c \tag{6}$$

where ρ_a is the aerosol density, U the superficial gas velocity, and d_c the diameter of the collector particle. The Stokes number can be interpreted as the ratio of the "stopping distance" for an aerosol particle (i.e. the distance within which the particle comes to rest when projected at velocity U into stagnant gas of viscosity μ) to the diameter of the collector. For given aerosol, collector and gas velocity, the Stokes number is sensitive to temperature through the gas viscosity, but relatively insensitive to pressure.

79

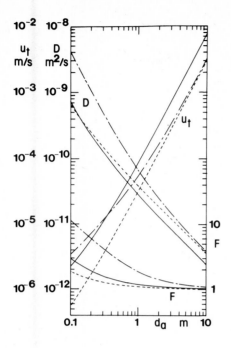

Fig.3. Slip correction factor (F),
Brownian diffusivity (D), and terminal
velocity (u_t) for spherical particles
in air:
——————— 300K, 1 bar;
λ = 0.069µm, μ = 1.85 x 10^{-5}Nam^{-2}
—·——·——1100K, 1 bar;
λ = 0.32 µm, μ = 4.55 x 10^{-5}Nsm^{-2}
— — — —1100K, 10 bar;
λ = 0.032µm, μ = 4.40 x 10^{-5}Nam^{-2}

In addition to the effect of inertia, the trajectory of an aerosol particle deviates from the gas streamlines due to settling under the influence of gravity (see Fig.2c). The effect of gravitational settling on collection is normally expressed in terms of the ratio of the terminal velocity of the particle in still gas, u_t, to the superficial velocity through the filter bed:

$$N_G = u_t/U \qquad (7)$$

The terminal settling velocity follows from Stokes' law as (ref.3)

$$u_t = Fgd_a^2(\rho_a - \rho_g)/18\mu \qquad (8)$$

so that the gravitational settling parameter can be written

$$N_G = Fgd_a^2(\rho_a - \rho_g)/18\mu U \qquad (9)$$

Values for the terminal settling velocity for particles of density 2500 kg/m^3 are shown in Fig.3. It may again be noted that u_t is sensitive to temperature but not to pressure.

In treating diffusional, inertial, and gravitational collection, an aerosol particle is essentially considered as a point mass. The fourth collection mechanism, direct interception, accounts for the finite size of a filtered aerosol particle. Even if the aerosol particle follows the gas streamlines faithfully, it will be collected if its centre passes within a distance $d_a/2$ from the surface of the collector (see Fig.2d). Collection by direct interception is usually correlated by the "interception parameter", which is simply the diameter ratio:

$$N_{DI} = d_a/d_c \qquad (10)$$

In addition to these four purely mechanical collection processes, filtration can be aided by electrophoretic effects; these are discussed in Section V. For completeness, two other processes should be noted. Aerosol particles will migrate in the presence of a concentration gradient in the gas. This effect is known as diffusiophoresis, and the migration velocity can be taken as the mean local mass velocity of the gas (ref.8). In principle, collection can be aided by diffusio-phoresis if the collector is also absorbing some component from the gas, but impractically high mass fluxes would be required for the effect to be perceptible in a granular filter. Particles also migrate in the presence of a temperature gradient, undergoing thermophoresis towards the lower temperature. In principle, thermophoretic collection could occur in a granular filter if the bed were maintained at a temperature below that of the entering gas, but this effect does not appear to have been reported.

II.2 Collection efficiency

Filter performance is conventionally expressed in terms of the collection efficiency of a single filter particle,

$$E = \frac{\text{Number of aerosol particles collected}}{\text{Number of particles in gas approaching collector}} \tag{11}$$

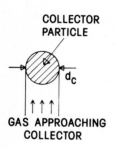

COLLECTOR PARTICLE

d_C

GAS APPROACHING COLLECTOR

Fig.4 Definition of collection efficiency for fixed bed.

The denominator of eqn.(11) requires careful definition; the "gas approaching collector" is universally based on the cross-sectional area of the collector ($\pi d_c^2/4$; see Fig.4), but agreement is not universal on the appropriate value to use for the approach gas velocity. Some authors (e.g. refs 9,10) use the interstitial gas velocity, but it is more appropriate (ref.11) to base the approach gas volume on the super-ficial gas velocity through the filter, U. With the latter definition, the volume of gas per unit time approaching the collector is $U\pi d_c^2/4$. If the number of aerosol particles per unit volume in the approaching gas is C, then the number of particles per unit time collected by the individual collector particle is

$$r = EUC\pi d_c^2/4 \tag{12}$$

Each of the mechanisms discussed in the preceding section contributes to the total collection efficiency, E. The total efficiency is commonly and conveniently taken as the sum of the efficiencies for the individual mechanisms, i.e.

$$E = E_D + E_I + E_G + E_{DI} \tag{13}$$

where the suffices refer respectively to diffusion, inertia, gravity, and direct interception. The approximation implied by eqn.(13) is valid if the individual efficiencies are all small or if one mechanism dominates (ref.12). Normally these conditions are met in granular filters, and eqn.(13) is therefore used here. In general, the total efficiency is greater than any of the individual efficiencies and smaller than their sum (ref.13).

TABLE 1

Efficiencies of individual filtration mechanisms in a fixed bed

Mechanism	Efficiency	Range of Validity	References
Brownian diffusion	$E_D = \dfrac{4.36}{\varepsilon}\left[\dfrac{D}{Ud_c}\right]^{2/3}$	$165 < Sc < 70,600$ $Re < 55$	14,16,17,18
Inertial deposition	$E_I = \left[\dfrac{St}{St+0.062\varepsilon}\right]^3$	$2 \times 10^{-3} < St < 2 \times 10^{-2}$ $Re_c < 130$	14
Gravitational settling	$E_G = 0.0375\sqrt{N_G}$	$0.05 < Re_c < 30$	10,14
Direct interception	$E_{DI} = 6.3N_{DI}^2\varepsilon^{-2.4}$	?	9

Table 1 summarises the most reliable correlations currently available for the efficiencies of the individual mechanisms. Apart from direct interception, the expressions in Table 1 are all well suported by experimental measurement (ref.14). Theoretical treatments are available (see ref.15), but their predictions do not always agree satisfactorily with the available data (10,11,14). The expression for E_D relies on the analogy between diffusional collection and mass transfer in packed beds at high values of the Schmidt number:

$$Sc = \mu/\rho_g D \qquad (14)$$

where D is the molecular or Brownian diffusivity. From eqn.(12), the efficiency of diffusional collection is related to the conventional mass transfer coefficient, k_D, by

$$E_D = 4k_D/U \qquad (15)$$

and the result in Table 1 is obtained from the correlation of Wilson and Geankoplis (ref.16) for mass transfer at high Sc. The expression for the efficiency of inertial deposition was obtained by Thambimuthu (ref.14) from correlation of all available data. The correlation for gravitational settling applies when the gas flows vertically upwards through the filter, the case of interest for fluidised beds. It was first proposed by Paretsky (ref.10), and is consistent with more recent data (ref.14). The expression for the efficiency of collection by direct interception was obtained from a theoretical analysis by Paretsky *et al.* (ref.9); the accuracy of this result has not been systematically tested, but it is consistent with available data (ref.14). Table 1 also indicates the range of validity claimed for each result, in terms of the groups describing filtration and of the Reynolds number of the collector particle:

$$Re_c = U\rho_g d_c/\mu \qquad\qquad (16)$$

In most cases, other mechanisms dominate outside the range of validity of each
correlation, but more data are needed to extend the range of inertial efficiency to
higher St and Re_c. The results in Table 1 all refer to roughly spherical collector
granules, and collection efficiencies may be higher on irregular particles (ref.14).

Figures 5 and 6 illustrate the relative importance of the four collection
mechanisms. The gas velocity for each figure corresponds to minimum fluidisation
of the collector, for reasons explained in Section IV. Generally diffusional
collection predominates for small aerosol particles and low gas velocities, while
for large aerosols and high velocity collection occurs almost entirely by inertial
impaction. Between these ranges, the collection efficiency passes through a

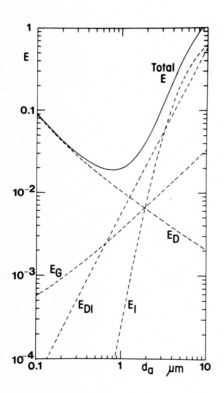

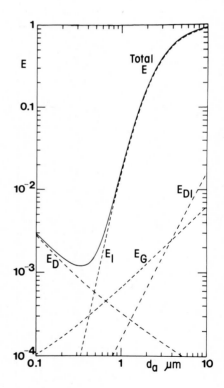

Fig.5 Effect of aerosol size on
collection efficiency, for 100μm
collector filtering air at 300K
and 1 bar.

U = 9.5mms^{-1}; ε = 0.4.

Fig.6 Effect of aerosol size on
collection efficiency, for 600μm
collector filtering air at 300K
and 1 bar.

U = 0.28ms^{-1}; ε = 0.4.

minimum. The corresponding aerosol diameter, usually termed the "most penetrating size" (ref.19), is a function of collector size and, for fixed beds, gas velocity (ref.7).

II.3 Overall penetration

The overall performance of a gas filter is usually described by the overall penetration:

$$f = \frac{\text{aerosol concentration in exit gas}}{\text{concentration in gas entering filter}} \tag{17}$$

or by the overall efficiency:

$$\eta = 1 - f \tag{18}$$

The following analysis to relate overall penetration to the efficiency of an individual collector is summarised from refs.(11) and (20), and assumes that E is constant through the filter. It therefore applies to deep-bed filtration in the stationary régime, and to a single size of aerosol. Since the analysis proceeds by treating the filter medium as a continuum, the bed must also be deep in the sense that the filter thickness is large compared to the diameter of an individual collector. Deep bed filtration of heterodisperse aerosols can be treated by considering each of the component sizes present, but the average local value of E then decreases through the filter as all but the most penetrating particles are removed (ref.7).

If the void fraction in the filter bed is ε, then the number of collector particles per unit volume is $6(1-\varepsilon)/\pi d_c^3$. The number of aerosol particles per unit time collected per unit bed volume is then

$$R = 6r(1-\varepsilon)/\pi d_c^3 = 3EUC(1-\varepsilon)/2d_c \tag{19}$$

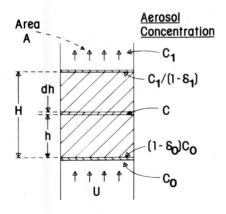

Fig.7 Aerosol filtration in a fixed bed (schematic).

where r is the rate of collection by a single granule, given by eqn.(12), and C is the local number concentration of aerosol in the interstitial gas. Now consider an aerosol-laden gas flowing through a fixed bed filter of cross-sectional area A and total depth H (Fig.7). If the concentration at distance h from the gas inlet is C, then the total rate of collection in the elemental depth dh is

$$ARdh = \left[3EUC(1-\varepsilon)/2d_c\right]A\,dh \tag{20}$$

An aerosol balance over this element then yields

$$\frac{dC}{dh} = -\frac{AR}{AU} = -\left[\frac{3E(1-\varepsilon)}{2d_c}\right] C \tag{21}$$

Thus the group

$$\kappa = 3E(1-\varepsilon)/2d_c \tag{22}$$

represents a rate constant for filtration; its significance is that an aerosol particle has probability κdh of being collected on passage through a depth dh of filter.

On entry to the filter bed, some anomalous collection may occur, especially if the bed is held in place by a retaining grid. If the aerosol concentration in the gas approaching the bed is C_o, and the fraction of aerosol collected on entry is δ_o, then the effective concentration entering the granular bed is $(1-\delta_o)C_o$. Solving eqn.(21) with this initial condition yields

$$C = (1-\delta_o)C_o \exp(-\kappa h) \tag{23}$$

Similarly, a fraction δ_1 of the aerosol approaching any retaining grid at the filter exit may be collected. The final exit concentration is then

$$C_1 = (1-\delta_o)(1-\delta_1)C_o \exp(-\kappa H) \tag{24}$$

or, in terms of penetration

$$f = C_1/C_o = (1-\delta_o)(1-\delta_1)\exp(-\kappa H) \tag{25}$$

Equation (25) enables determination of κ and hence E without interference from end effects (11,20). In the present context, it also gives one limit for the penetration through a fluidised filter (see Section III.5).

III. ANALYSIS OF FILTRATION IN FLUIDISED BEDS

III.1 General behaviour of fluidised beds

Before examining gas filtration in fluidised beds, it is necessary to summarise the essential hydrodynamic features of these devices. Further details can be obtained from one of the excellent texts available (e.g. refs. 21-23).

A fluidised bed is a mass of granular material supported by upward flow of a gas or liquid. The fluid enters the bed through a distributor whose primary function is to ensure uniform distribution of the fluidising fluid across the containing column. Consider a packed bed with a fluid passing upwards through it.

At low superficial velocities, the pressure drop across the bed is less than the weight of granular material per unit cross-sectional area; the granules are then supported partly by interparticle forces and partly by fluid drag. As the superficial velocity is increased, the pressure drop across the bed increases, so that the proportion of the bed weight supported by drag also increases. Eventually, at the velocity corresponding to "minimum fluidisation" or "incipient fluidisation", the pressure drop exactly counterbalances the bed weight so that the particles are entirely supported by the fluid. Above this velocity, the bed is said to be fluidised, and takes on many of the properties of a viscous liquid. For spherical particles, the superficial velocity at minimum fluidisation, U_{mf}, can be estimated from a result derived from Ergun's equation for pressure drop through a packed bed (ref.24):

$$Ga = 150 \frac{(1-\varepsilon_{mf})}{\varepsilon_{mf}^3} Re_{mf} + 1.75 \frac{1}{\varepsilon_{mf}^3} Re_{mf}^2 \qquad (26)$$

where ε_{mf} is the void fraction in the bed at minimum fluidisation, Re_{mf} is the particle Reynolds number at minimum fluidisation:

$$Re_{mf} = U_{mf}\rho_g d_c/\mu \qquad (27)$$

and Ga is a dimensionless property group sometimes known as the "Galileo number":

$$Ga = \rho_g(\rho_c-\rho_g)gd_c^3/\mu^2 \qquad (28)$$

where ρ_c is the density of a bed particle. For spherical particles, ε_{mf} is typically 0.4, and eqn.(26) then becomes (22)

$$Re_{mf} = 25.7\left[\sqrt{1 + 5.53 \times 10^{-5}Ga} - 1\right] \qquad (29)$$

For irregular particles, estimation of U_{mf} is complicated by uncertainty over both the effect of shape on drag and over the value of ε_{mf}. Reasonable estimates can usually be obtained from a correlation due to Wen and Yu (ref.25):

$$Re_{mf} = 33.7\left[\sqrt{1 + 3.59 \times 10^{-5}Ga} - 1\right] \qquad (30)$$

Particles of interest as collection media for fluidised filters are in Group B of the classification proposed by Geldart (ref.26). One characteristic of such particles is that, as soon as the superficial velocity exceeds U_{mf}, the fluidising gas divides into two distinct streams. Part of the gas continues to percolate between the bed particles to maintain fluidisation, while the remainder forms large rising voids, known as "bubbles" since they show many of the features of large bubbles in viscous liquids (ref.21). The gas in the bubble phase is not in direct contact with the bed particles, so that aerosol in the bubble gas is not directly available for filtration. The continuous "phase" of fluidised particles between the dispersed bubbles is known as the "particulate", "dense", or "emulsion" phase.

The division of gas flow between the particulate and bubble phases is still a matter for debate. Much work on gas-fluidised beds is based on the two-phase theory of fluidisation, proposed by Toomey and Johnstone (ref.27) and developed by Davidson and Harrison (ref.21). According to the two-phase theory, the voidage in the particulate phase remains constant at ε_{mf} and the flow rate through the particulate phase is equal to the flow rate at minimum fluidisation. The simple two-phase theory is open to theoretical objections (ref.28), and also leads to over-estimation of bubble flow rate (ref.29). The theoretical objections concern the interstitial flow in the particulate phase, and are therefore critical for fluidised filters where this flow determines the efficiency of filtration. The "modified two-phase theory" first proposed by Lockett *et al.* (ref.30) gives improved but still not exact estimates of bubble flow rate (ref.29). However, since it is self-consistent in treating flow in the particulate phase, it provides a suitable starting point for the analysis of fluidised filtration. Figure 8 shows a schematic section through a bubbling bed. At any section AA', the time-averaged fraction of the cross-sectional area occupied by the bubble phase is denoted by ε_B. Under steady operating conditions, there is no net transfer of bed particles across the section. According to the modified two-phase model, the flow of fluidising gas within the average area $A(1-\varepsilon_B)$ occupied by particulate phase is equal to its value at minimum fluidisation, i.e. $U_{mf}A(1-\varepsilon_B)$. The balance of the gas flow, i.e. $A\left[U-U_{mf}(1-\varepsilon_B)\right]$, crosses the section in the bubble phase. In the following analysis, it is convenient to denote the fraction of fluidising gas crossing any section within the bubble phase by β; i.e.

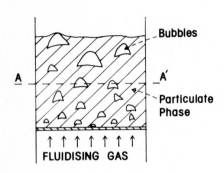

Fig.8 Schematic section through bubbling fluidised bed.

$$\beta = 1 - U_{mf}(1-\varepsilon_B)/U \tag{31}$$

Clearly $(1-\beta)$ represents the fraction of the total gas flow passing through the particulate phase.

III.2 <u>Transfer between bubbles and particulate phase</u>

There is one important respect in which bubbles in fluidised beds differ from bubbles in liquids: because the bubble boundary is an interface between a gas void and a porous medium, it is permeable to gas. As a result, there is a net flow of gas, shown schematically in Fig.9, into the base of the bubble and out

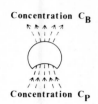

Concentration C_B

Concentration C_P

Fig.9 Gas flow
through bubble in
fluidised bed
(schematic)

through the roof, the mean velocity within the bubble being of
order $3U_{mf}$ (ref.21). This "throughflow" convects aerosol
particles between the bubble and the particulate phase. The
gas within the bubble can be taken as fully mixed (ref.21),
with concentration C_B of aerosol. If the aerosol concentration
in the gas entering the bubble from the surrounding particulate
phase is C_P, then the net rate of convective transfer from
bubble to particulate phase is $q(C_B-C_P)$, where q is the
volumetric gas flowrate through the bubble.

In addition to this convective transfer, aerosol can also
be exchanged by diffusive processes. It is convenient to
define a mass transfer coefficient, k_B, so that the rate of diffusional transfer
is $k_B S(C_B-C_P)$ where S is the surface area of the bubble. The total net rate of
interphase transfer per unit bubble volume is then

$$J \quad = \quad \frac{(q + k_B S)}{V} \quad (C_B-C_P) \tag{32}$$

where V is the bubble volume. Davidson and Harrison (ref.21) have given expressions for q and $k_B S$ which serve to indicate the relative magnitudes of convective
and diffusive transfer:

$$q/V \quad = \quad 9U_{mf}/2d_B \tag{33}$$

$$k_B S/V \quad = \quad 5.85g^{\frac{1}{4}}D^{\frac{1}{2}}d_B^{-5/4} \tag{34}$$

where d_B is the volume-equivalent bubble diameter. Table 2 gives typical values
for q/V and $k_B S/V$, evaluated from eqns.(33) and (34) using values for Brownian
diffusivity, D, from Fig.3 and assuming a bubble diameter of 0.01m. Since the
ratio of diffusive to convective transfer is proportional to $d_B^{-\frac{1}{4}}$, the relative
importance of the two mechanisms is insensitive to bubble size. Clearly diffusive
transfer is insignificant by comparison with convective transfer, except for aerosol
particles much smaller than the micron range to which fluidised filters are
considered applicable. Since interphase transfer is dominated by convection, the
transfer rate is independent of the aerosol size. This conclusion is of considerable importance in interpreting the performance of fluidised filters.

Although eqns.(33) and (34) serve to indicate the relative magnitudes of the two
transfer mechanisms, they can only give approximate values for the transfer rate.
Bubble coalescence in particular increases interphase transfer. Although the
significance of coalescence has yet to be quantified, its effect is to increase
convective exchange (ref.31) confirming the predominance of this mechanism. In
the analysis of fluidised filtration which follows, the net rate of interphase
transfer will be taken as $K(C_B-C_P)$ per unit volume of bubble phase, with K resulting

TABLE 2

Typical values for convective and diffusive transfer of aerosol from bubble to particulate phase. Gas properties correspond to air; bubble diameter, d_B, taken as 0.01m.

Collector diameter, d_c	Conditions	Aerosol diameter, d_a (μm)	q/v (s^{-1})	$k_B S/V$ (s^{-1})
100μm	300K, 1 bar U_{mf} = 9.5 mm s^{-1}	0.1	4.3	0.087
		0.3	4.3	0.037
		1	4.3	0.017
		3	4.3	0.0095
	1100K, 1 bar U_{mf} = 3.8 mm s^{-1}	0.1	1.7	0.21
		0.3	1.7	0.073
		1	1.7	0.027
		3	1.7	0.013
	1100K, 10 bar U_{mf} = 3.8 mm s^{-1}	0.1	1.7	0.084
		0.3	1.7	0.040
		1	1.7	0.020
		3	1.7	0.011
500μm	300K 1 bar U_{mf} = 0.28 m s^{-1}	0.1	130	0.087
		0.3	130	0.037
		1	130	0.017
		3	130	0.0095
	1100K 1 bar U_{mf} = 0.14 m s^{-1}	0.1	61	0.21
		0.3	61	0.073
		1	61	0.027
		3	61	0 013
	1100K 10 bar U_{mf} = 0.12 m s^{-1}	0.1	56	0 084
		0.3	56	0.040
		1	56	0.020
		3	56	0.011

entirely from convection and independent of d_a. As in the analysis of chemical reaction in fluidised beds first developed by Orcutt *et al.* (ref.32), it is assumed that aerosol convected from the bubble phase is distributed through the particulate phase either at the horizontal level in question (plug flow case; Section III.3) or throughout the bed (fully mixed case: Section III.4). More complex models of reaction in fluidised beds have been proposed (e.g. ref.23), to account for the fact that gas flowing through a bubble may recirculate within a limited "cloud" region surrounding the bubble. However, it is not clear either that the "cloud" exists as a well-defined region in most freely-bubbling beds (refs. 31, 33) or that the additional complexity leads to an improved description of fluidised bed reactors (ref.34). The simpler approach developed from the reactor model of Orcutt *et al.* (ref.32) is therefore used in this review.

Consider now a fluidised bed of cross-sectional area A, through which the super-ficial gas velocity is U. At height h above the distributor the average concen-tration of aerosol is C_B in the bubble phase and C_p in the interstitial gas in the

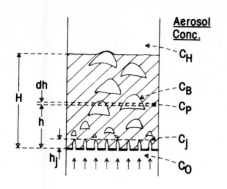

Fig.10 Fluidised bed filter
(schematic)

particulate phase. The flow rate of
bubble phase gas across the section is
βAU (see Section III.1). Within the bed
element between planes at heights h and
(h+dh) above the distributor, shown
schematically in Fig.10, the volume of
bubble phase is $\varepsilon_B Adh$. The net transfer
of aerosol from bubbles to particulate
phase within this element is then
$K(C_B-C_P)\varepsilon_B Adh$. A balance on the aerosol
in the bubble phase then yields

$$\frac{dC_B}{dh} = \frac{-K\varepsilon_B}{\beta U}(C_B-C_P) \tag{35}$$

Since transfer occurs entirely by convective "throughflow", with aerosol at concentration C_B replaced by aerosol at concentration C_P, eqn.(35) implies that the fraction of the original bubble phase gas remaining within that phase after passage through a depth y of the bed is ϕ, where

$$\ln\phi = -\int_o^y \frac{K\varepsilon_B}{\beta U} dh \tag{36}$$

Making the simplifying assumption that K, ε_B and β can be treated as constant at their average values in the bed,

$$\phi = \exp\left[-y K\varepsilon_B/\beta U\right] \tag{37}$$

Bubbles are formed at some distance h_j above the distributor, where the gas jets formed at orifices in the distributor plate break down into a stream of bubbles (ref.35). Thus bubbles reaching the surface of the bed a distance H above the distributor have passed through a depth $(H-h_j)$. From eqn.(37), the fraction of the initial bubble phase gas remaining untransferred at the surface is then

$$\phi = \exp\left[-(H-h_j)K\varepsilon_B/\beta U\right] \tag{38}$$

The exponent in eqn.(38) is a dimensionless group commonly termed the "crossflow factor":

$$X = (H-h_j)K\varepsilon_B/\beta U \tag{39}$$

The crossflow factor may be interpreted as the number of bubble volumes of gas flowing through a bubble during its rise through the bed.

III.3 Filtration with plug flow in particulate phase

The following analysis of filtration in a fluidised bed is a modification of the model developed by Doganoglu *et al.* (ref.36). The difference lies in the definition of single particle collection efficiency: the definition adopted here facilitates comparison with filtration in fixed beds. We consider first the case where the gas passing through the particulate phase is in plug flow. A single collector particle in a fluidised bed is shown schematically in Fig.11. There are now two processes causing aerosol to approach the collector: convection by the interstitial

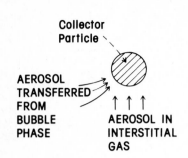

Collector
Particle

AEROSOL
TRANSFERRED
FROM
BUBBLE
PHASE

AEROSOL IN
INTERSTITIAL
GAS

Fig.11 Definition of collection efficiency for fluidised bed

fluidising gas, and transfer from the bubble phase. Assuming that the void fraction in the particulate phase is ε_{mf}, the volume of particulate phase associated with a single collector is $\pi d_c^3/6(1-\varepsilon_{mf})$. Thus a single layer of collector particles in the particulate phase is equivalent to a bed depth of

$$\left[\pi d_c^3/6(1-\varepsilon_{mf})\right]^{1/3} = \alpha d_c \qquad (40)$$

where

$$\alpha = \left[\pi/6(1-\varepsilon_{mf})\right]^{1/3} = 0.96 \text{ for } \varepsilon_{mf} = 0.4 \qquad (41)$$

From eqn.(37), the fraction of bubble phase gas transferred to the particulate phase on passing through this depth αd_c is $(1-e^{-Y})$ where

$$Y = \alpha d_c K \varepsilon_B/\beta U = \alpha d_c X/(H-h_j) \qquad (42)$$

The number of aerosol particles per unit time carried to the section in question by the bubble gas is $A\beta UC_B$. Therefore the number of particles per unit time presented to the single layer of collector particles by transfer from the bubble phase is $(1-e^{-Y})A\beta UC_B$. Within the particulate phase, the gas flow rate is $(1-\beta)AU$ so that the interstitial gas presents aerosol to the layer of collector at a rate $(1-\beta)AUC_P$. Thus the total rate at which aerosol is presented to the depth αd_c of particulate phase is

$$n = (1-\beta)AUC_P + (1-e^{-Y})A\beta UC_B \qquad (43)$$

By analogy with the treatment of fixed bed filters in Section II.3, it is convenient to define a filtration rate constant κ such that the fraction of these aerosol particles collected is $\kappa \alpha d_c$. The total collection rate in the depth αd_c of bed is then

$$\kappa \alpha d_c n = \kappa \alpha d_c \left[(1-\beta)AUC_P + (1-e^{-Y})A\beta UC_B \right] \qquad (44)$$

from eqn.(43). The depth αd_c of fluidised bed contains a volume $(1-\varepsilon_B)A\alpha d_c$ of particulate phase. Therefore the collection rate per unit particulate phase volume is

$$R = \kappa n/A(1-\varepsilon_B)$$

$$= \frac{\kappa}{(1-\varepsilon_B)} \left[(1-\beta)UC_P + (1-e^{-Y})\beta UC_B \right] \qquad (45)$$

From eqn.(31), derived from the modified two-phase theory of fluidisation

$$(1-\beta)U = (1-\varepsilon_B)U_{mf} \qquad (46)$$

Eqn.(45) then becomes

$$R = \kappa \left[U_{mf}C_P + (1-e^{-Y})\beta UC_B/(1-\varepsilon_B) \right] \qquad (47)$$

The first term in eqn.(47) is the filtration rate in a packed bed at superficial velocity U_{mf} (*cf*. eqns.(19) and (22)), while the second term describes collection of aerosol transferred from the bubble phase. If the single particle collection efficiency, E, is defined in the same way as for fixed bed filters, then E and κ are related by eqn.(22) which may be written for a fluidised filter as

$$\kappa = 3E(1-\varepsilon_{mf})/2d_c \qquad (48)$$

In terms of E, eqn.(47) then becomes

$$R = \frac{3E(1-\varepsilon_{mf})}{2d_c} \left[U_{mf}C_P + \frac{(1-e^{-Y})}{(1-\varepsilon_B)} \beta UC_B \right] \qquad (49)$$

As in the treatment of fixed beds in Section II.3, the particulate phase has been treated here as a continuum, so that eqns.(47) to (49) only apply if $(H-h_j)$ is large by comparison with d_c.

Consider now an infinitesimal element of depth dh of the fluidised bed in Fig.10. The volume of particulate phase within this element is $(1-\varepsilon_B)$dh. A total aerosol balance over the element yields

$$\beta U \frac{dC_B}{dh} + (1-\beta)U \frac{dC_P}{dh} + R(1-\varepsilon_B) = 0 \qquad (50)$$

where the first two terms represent the change with h of aerosol flow in the bubble and particulate phases and the third term represents collection in the particulate phase. Substituting for R from eqn.(45) yields

$$\beta \frac{dC_B}{dh} + (1-\beta)\frac{dC_P}{dh} + (1-\beta)\kappa C_P + (1-e^{-Y})\beta \kappa C_B = 0 \qquad (51)$$

Eqn.(51) is to be solved simultaneously with eqn.(35),which can be written

$$\frac{dC_B}{dh} + \frac{X(C_B-C_P)}{(H-h_j)} = 0 \tag{52}$$

where X is the crossflow factor defined by eqn.(39),

As for the analysis of packed bed filtration in Section II.3, the initial conditions for eqns.(50) and (51) must allow for anomalous collection at entry to the filter. Eqns.(50) and (51) apply to the bubbling bed above the jet region, i.e. $h > h_j$. If the aerosol concentration is C_o in the gas approaching the filter bed and C_j at the end of the jet region, then the penetration through the jets is defined as

$$f_j = C_j/C_o \tag{53}$$

The significance of f_j is discussed in Section IV.2. Assuming that gas enters both bubble and particulate phases with concentration C_j, the initial conditions are

$$C_B = C_P = C_j \quad \text{at} \quad h = h_j \tag{54}$$

It is convenient to introduce the dimensionless groups

$$z = (h-h_j)/(H-h_j) \tag{55}$$

$$k = \kappa(H-h_j) = 3E(1-\epsilon_{mf})(H-h_j)/2d_c \tag{56}$$

Eqns.(51), (52) and (54) then become

$$\beta\frac{dC_B}{dz} + (1-\beta)\frac{dC_P}{dz} + (1-\beta)kC_P + (1-e^{-Y})\beta kC_B = 0 \tag{57}$$

$$\frac{dC_B}{dz} + X(C_B-C_P) = 0 \tag{58}$$

$$C_B = C_P = C_j \quad \text{at} \quad z = 0 \tag{59}$$

The gas leaving the filter bed is a mixture of the gas streams from the bubble and particulate phases, so that the exit concentration is given by

$$C_H = \beta(C_B)_{z=1} + (1-\beta)(C_P)_{z=1} \tag{60}$$

Solving eqns.(57) to (59) and inserting the results into eqn.(60) yields

$$C_H = \frac{C_j}{(m_2-m_1)}\left[m_2 e^{-m_1}\left\{1 - \frac{(1-\beta)m_1}{X}\right\} - m_1 e^{-m_2}\left\{1 - \frac{(1-\beta)m_2}{X}\right\}\right] \tag{61}$$

where m_1, m_2 are the roots of

$$(1-\beta)m^2 - \left[X + (1-\beta)k\right]m + kX(1-\beta e^{-Y}) = 0 \tag{62}$$

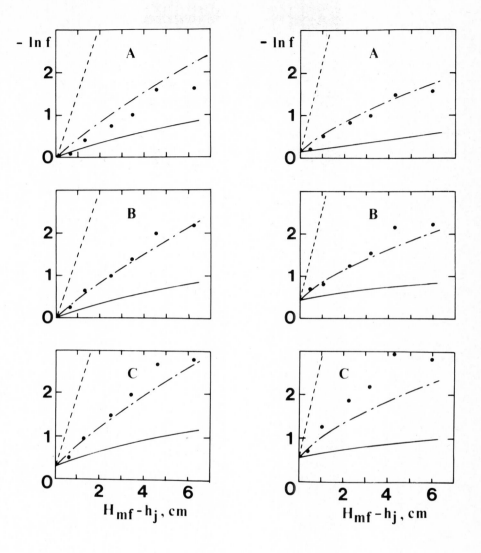

Fig.21. Overall penetration: Fig.22. Overall penetration:

$U = 0.06$ m s^{-1} $U = 0.13$ m s^{-1}

Data of Doganoglu (ref. 20) for di-octyl phthalate (liquid)aerosol in
air at ambient temperature and pressure in beds of glass ballotini

$\rho_c = 2440$ kg m^{-3}; $d_c = 108\mu$m; $U_{mf} = 0.02$ m s^{-1}; $d_{or} = 1.23$ mm

d_a: A - 0.72μm; B - 0.9μm; C - 1.15μm.

                – – – – – – – – – $X \to \infty$

                ———————— X from ref. 37

                –·—·—·—·—·—· X' from eqn.(94)

aerosol diameter, d_a, leads to increased jet collection. Figures 21 and 22 refer to data which differ only in the gas superficial velocity, U, and show that jet collection increases with increasing U. It is also clear (Fig.20) that, at high gas velocity, jet collection accounts for the greater part of the overall collection. These effects are discussed further in Section IV.2.

The variation of (-lnf) with H_{mf} for $H_{mf} > h_j$ results from processes occurring in the bubbling part of the filter, and is discussed in Section IV.3 along with the question of which collection mechanisms dominate. In general, increasing gas velocity increases the fraction of gas passing through the bubble phase, β, and also tends to decrease the crossflow factor, X (ref.37). It was shown in Section III.5 that both these effects increase "bypassing" of aerosol, and hence increase f_b. Thus the effects of gas velocity on f_j and f_b are in opposition, and the overall effect of gas velocity depends on which mode of collection dominates. Figure 23 shows data reported by McCarthy *et al.* (ref.45) and Doganoglu (ref.20). At velocities close to minimum fluidisation there is little collection in the jets (i.e. $f_j \approx 1$) so that overall penetration is determined by f_b and increases as the velocity is raised. At higher velocities, jet collection becomes significant; the overall penetration passes through a maximum and then decreases with further increase in velocity. The position of the maximum depends on the characteristics of the distributor and collector. The majority of experimental studies have been carried out at velocities well above the maximum, so that most workers have reported penetration smaller than the values in Fig.23 and decreasing with increasing U.

IV.2 Collection in distributor jets

In conventional fluidised beds, the distributor plate is required to distribute the fluidising gas uniformly across the bed. For fluidised filters, the distributor must be designed to meet two further requirements. To avoid blockage, it must not collect appreciable quantities of the challenging dust or mist. To aid filtration, it must be designed to give high collection efficiency in the gas jets formed at the distributor. Most experimental studies have used distributors of the multiple orifice type, consisting of a flat plate drilled with a regular array of circular holes. Collection of aerosol by the distributor itself is reduced if the orifices are countersunk on the lower (upstream) face, as shown in Fig.24. If the half-angle of the countersunk hole, θ, is less than about 30°, extended operation without appreciable blockage can be attained (ref.38).

When the fluidising gas is introduced through this type of distributor, an array of gas jets forms above the orifices. Although the flow in the jets may not be steady (ref.51), it is sufficient to regard the jets as regions in which the average gas velocity is high and the concentration of fluidised solids is low by comparison with the bed above the gas entry region. Available evidence (ref.52)

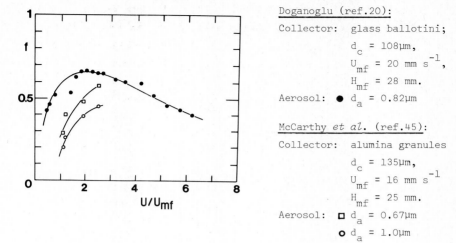

Doganoglu (ref.20):

Collector: glass ballotini;

$d_c = 108\mu m$,

$U_{mf} = 20$ mm s^{-1},

$H_{mf} = 28$ mm.

Aerosol: ● $d_a = 0.82\mu m$

McCarthy *et al.* (ref.45):

Collector: alumina granules

$d_c = 135\mu m$,

$U_{mf} = 16$ mm s^{-1},

$H_{mf} = 25$ mm.

Aerosol: □ $d_a = 0.67\mu m$

○ $d_a = 1.0\mu m$

Fig.23. Overall penetration of di-octyl phthalate (liquid) aerosols in air at ambient temperature and pressure through shallow fluidised beds.

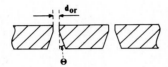

Fig.24. Schematic section through distributor plate for fluidised filter, after Ghadiri (ref.38).

suggests that, over the range of conditions of interest for fluidised filters, the correlation due to Merry (ref.35) gives reasonable estimates for the length of the jet, h_j. This correlation has frequently been employed in analysing data on fluidised filtration (refs.14,36,38), and was used to estimate h_j in Figs.20-22.

The processes occurring in the jets are shown schematically in Fig.25, following the model developed by Massimilla and his co-workers to describe momentum, heat and mass transfer in the jet region (e.g. refs.53,54). Collector particles are entrained into the jet from the partially-fluidised zone surrounding the orifice, and are accelerated upwards by the high velocity gas. Since the mean gas velocity through the distribution orifices is typically of order 10 to 100 m/s in a fluidised filter (ref.36), the jet must represent a region in which the slip velocity is high. Under such circumstances, the dominant collection mechanism is likely to be inertial impaction (*cf*. Section II). It was noted in Section IV.1

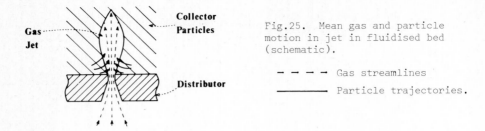

Fig.25. Mean gas and particle motion in jet in fluidised bed (schematic).

- - - - Gas streamlines
———————— Particle trajectories.

that jet collection increases with gas velocity and aerosol diameter, consistent with an inertial process.

Thambimuthu (ref.14) has analysed the available data on jet collection, inferred from plots similar to Figs.20 to 22. In addition to the effects of gas velocity and aerosol size, the penetration through the jet region decreases with increasing collector diameter, d_c, and decreasing orifice diameter, d_{or} (see Fig.24); both these trends correspond to increasing concentration of collector particles in the gas jet (ref.54). Pending more detailed analysis of the processes occurring in the jets, the data have been correlated in terms of the group

$$N_{or} = St_{or}(N_{uc})^{1/3} \tag{81}$$

where St_{or} is a Stokes number based on the mean gas velocity through the distributor orifice:

$$St_{or} = F\rho_a d_a^2 U_{or}/9\mu d_c \tag{82}$$

and $(N_{uc})^{1/3}$ is a dimensionless terminal velocity for the collector particles:

$$(N_{uc})^{1/3} = \left(\frac{3\rho_g^2}{4\rho_c g\mu}\right)^{1/3} u_{tc} \tag{83}$$

where ρ_c is the density of the collector and u_{tc} is the terminal velocity of a single collector granule in the gas at conditions corresponding to entry to the filter. The group St_{or} was used as independent variable in an earlier correlation (ref.36), and describes inertial impaction on an individual collector at slip velocity U_{or} (cf. eqn.6). For orifices of the general form shown in Fig.24, the discharge coefficient is close to unity and U_{or} can therefore be evaluated as the gas superficial velocity divided by the fractional open area in the distributor. If the orifice discharge coefficient departs significantly from unity, St_{or} should be based on the mean velocity at the *vena contracta* (ref.50). The group N_{uc} has been introduced to take some account of the increase in collector concentration and hence collection efficiency which result from increasing the

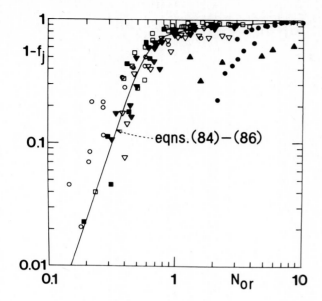

TABLE 4

Key to Fig.26

Ref.	Symbol	d_c (μm)	d_{or} (mm)
14	▽ ▼ □ ■	165 306 532 1100	0.6, 1.2
20	○ ●	108 600	1.22
43	▲	425	0.79

Fig.26. Empirical correlation of jet collection in beds of spherical collector, after Thambimuthu (ref.14).

density or diameter of the collector (ref.14). The terminal velocity, u_{tc}, can be estimated from the correlations given by Clift *et al.* (ref.3).

Figure 26 shows that the group N_{or} serves to reduce most of the available measurements on jet collection with spherical collectors (glass ballotini). The correlation is given by:

$$N_{or} < 0.4 : \quad f_j = 1 - 3.1 \, N_{or}^3 \tag{84}$$

$$0.4 < N_{or} < 0.8 : \quad f_j = 1.749 - 2.62 \, N_{or} + 0.345 \, N_{or}^2 + 0.715 \, N_{or}^3 \tag{85}$$

$$N_{or} > 0.8 : \quad f_j = 1 - \left(\frac{N_{or}}{N_{or} + 0.078} \right)^3 \tag{86}$$

The critical range of N_{or} is between 0.1 and 1.0: for $N_{or} < 0.1$ there is virtually no collection in the jet, while little further collection is achieved by increasing N_{or} above about unity.

Tan (50) has recently determined f_j for aerosols of di-ethyl hexyl sebacate (liquid) in beds of ballotini and sand; the data are shown in Fig.27. For spherical ballotini, the values of f_j again agree closely with the correlation of Fig.26. However, there is consistently less jet collection when irregular sand

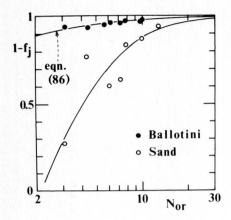

Fig.27. Jet collection of di-
ethyl hexyl sebacate aerosols
in beds of glass ballotini and
silica sand, after Tan (ref.50):

d_c = 461µm, d_{or} = 1.5mm.

is used as the collection medium. The sand was less free-flowing than ballotini:
the angle of repose was 6 to $10°$ larger than that for ballotini ($24° \pm 1°$).
Entrainment of collector into the jets is therefore slower than for ballotini,
leading to lower concentration of collector in the jets (ref.54) and hence reducing
collection.

IV.3 <u>Collection in bubbling part of bed</u>

The experimental data show that, for shallow beds, (-lnf) is approximately a
linear function of bed depth (see refs.14, 36 and 43 and Figs.20 to 22). The
analysis in Section III can be used to interpret this relationship. The majority
of the data correspond to values of U/U_{mf} less than 5, so that β is typically less
than 0.8 and the plug flow case is appropriate (see Section III.5). From Table 3,
the overall penetration for the limiting case X→∞, Y→∞ is

$$f = f_j e^{-k} \tag{87}$$

Hence, from the definition of k in eqn.(56),

$$-\ln f = -\ln f_j + \kappa(H-h_j) = -\ln f_j + \frac{3E(1-\epsilon_{mf})}{2d_c}(H-h_j) \tag{88}$$

Depths of fluidised beds are commonly reported as the corresponding depth at
minimum fluidisation conditions, H_{mf}. From the two-phase model outlined in
Section III.1,

$$H = H_{mf}/(1-\bar{\epsilon}_B) \tag{89}$$

where $\bar{\epsilon}_B$ is the average fraction of the whole bed occupied by bubbles. Eqn.(88)

then becomes

$$-\ln(f/f_j) = \frac{\kappa\left[H_{mf} - h_j(1-\bar{\epsilon}_B)\right]}{(1-\bar{\epsilon}_B)} \qquad (90)$$

Since h_j is typically much smaller than H_{mf}, the factor $(1-\bar{\epsilon}_B)$ in the numerator of eqn.(90) can be neglected, giving

$$-\ln(f/f_j) \approx \kappa(H_{mf}-h_j)/(1-\bar{\epsilon}_B) \qquad (91)$$

Thus if penetration through the bubbling part of the bed is limited by collection in the particulate phase (i.e. $X\to\infty$, $Y\to\infty$) and if $\bar{\epsilon}_B$ is small or weakly dependent on bed depth, then $(-\ln f)$ is a linear function of $(H_{mf}-h_j)$, as shown by the $X\to\infty$ curves in Figs.21 and 22. At the opposite limit $k\to\infty$, where penetration is controlled by interphase transfer, the corresponding result from Table 3 is

$$f = f_j \beta e^{-X} \qquad (92)$$

Using the definition of X in eqn.(39) and proceeding as in the development of eqn.(91),

$$-\ln(f/f_j) \approx \ln\beta - \frac{K\bar{\epsilon}_B(H_{mf}-h_j)}{(1-\bar{\epsilon}_B)\beta U} \qquad (93)$$

Thus, if K and $\bar{\epsilon}_B$ are weakly dependent on bed depth, $(-\ln f)$ is again a linear function of $(H_{mf}-h_j)$. Since the two limiting cases give approximately linear dependence, it is to be expected that the relationship between $(-\ln f)$ and $(H_{mf}-h_j)$ is also approximately linear for intermediate cases; Figs.20 to 22 demonstrate such dependence.

It was shown in Section III that, in the limit $k\to\infty$, penetration through the bubbling part of the bed is independent of aerosol size, but that this limit should not normally be approached. The data all show f_b dependent on d_a, and thus confirm this conclusion. However, the limit of rapid interphase transfer can be approached, and such data can be used to infer the collection rate constant, κ, and single-particle collection efficiency, E, in the particulate phase (ref.36). From eqn.(91), when $\bar{\epsilon}_B$ is small, the gradient of the relationship between $(-\ln f)$ and $(H_{mf}-h_j)$ gives an estimate for κ, i.e. $3E(1-\epsilon_{mf})/2d_c$. If interphase transfer has a secondary effect on penetration, the value of κ inferred in this way will be too low. On the other hand, if $\bar{\epsilon}_B$ is perceptible, the value will be too high (cf. eqn.(91)). Since the two effects act in opposition, this approach to estimating collection efficiency represents a reasonable compromise. The effect of gas velocity provides a convenient test of whether such conditions obtain. If penetration is affected by interphase transfer, then f_b increases with increasing U (see Section III.5). On the other hand, if interphase transfer is rapid, then U only affects f_j; the effect on plots of $(-\ln f)$ as a function of $(H_{mf}-h_j)$ is then to change the intercept at $(H_{mf}-h_j) = 0$ but to have essentially no effect on the gradient of the relationship for $H_{mf} > h_j$. It was also shown in Section III.5 that these conditions are most likely to be met for relatively "shallow" beds (i.e. low γ). Many measurements, such as the data for $d_a = 1.35\mu m$ in Fig.20, show this effect, departing from the linear relationship for $(H_{mf}-h_j)$ greater than about 50mm as interphase transfer starts to affect penetration (refs.14 and 36).

Thambimuthu (ref.14) examined the available measurements to identify data in which f_b appeared to be controlled by collection rather than interphase transfer. Linear regression was used to relate $(-\ln f)$ to $(H_{mf}-h_j)$, and the measurements taken to be controlled by collection rate if the gradient was independent of gas velocity. His own data include measurements in which the distributor characteristics were changed, but the aerosol size and superficial gas velocity were kept constant. For these results a further test was applied: the measurements can be taken as free from the influence of interphase transfer if the gradient of the regression line is not significantly affected by the distributor, since distributor characteristics affect the bubble size and hence X (see Section III.2). For data meeting these tests, the gradient was interpreted as an estimate for κ.

According to the two-phase model of fluidisation, conditions in the particulate phase of the fluidised bed should correspond to conditions in a fixed bed with void fraction ε_{mf} and superficial gas velocity U_{mf}. It is therefore of interest to compare the values of κ determined for fluidised filters with the values expected for filtration in a fixed bed under corresponding conditions, estimated by eqns.(13) and (22) using the expressions for the efficiencies of individual mechanisms summarised in Table 1. Figure 28 shows this comparison for one set of data. As required, the data show no systematic variation of κ with U/U_{mf} or distributor characteristics. The values scatter about the value for the equivalent fixed bed. Figure 29 shows the comparison with fixed bed values for all Thambimuthu's estimates for κ in fluidised beds. Each point in Fig.29 is an average over a range of gas velocities and, in some cases, distributor characteristics; the number of individual estimates averaged in this way is given in Table 6. The estimates scatter widely, although the scatter is attributable to small departures in void fraction and interstitial gas velocity from the values at minimum fluidisation (ref.14). However, the values scatter about the line representing equality with the collection rate in the corresponding fixed bed.

Thus it appears that the correlations for single-particle collection efficiency in fixed beds, summarised in Section II.2, can be applied to estimate the collection efficiency in the particulate phase of a fluidised bed. Moreover, Ghadiri's measurements on filtration of liquid aerosols at $800^{\circ}C$ (ref.38)

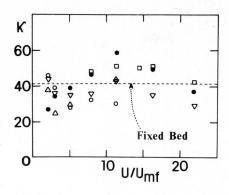

TABLE 5

Key to Fig.28

Symbol	Distributor	
	d_{or} (mm)	% open area
○	0.6	0.5
△	1.2	0.5
▽	0.6	1.5
□	1.2	1.5
●	0.6	3.0

Fig.28. Values of collection rate constant, κ, in fluidised bed of 165µm ballotini, after Thambimuthu (ref.14); d_a = 1.1µm.

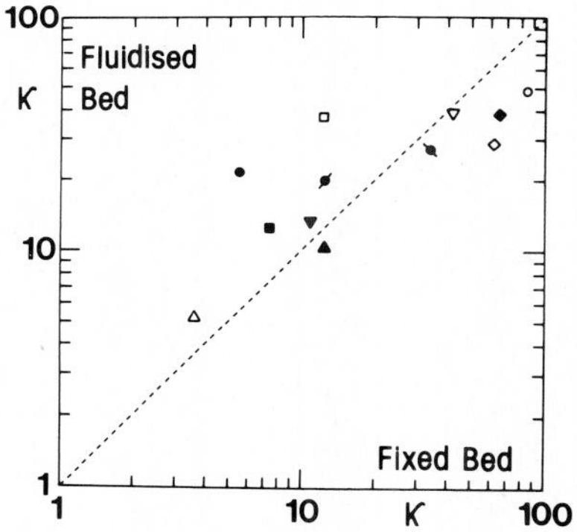

TABLE 6

Key to Fig.29

Ref.	Symbol	d_a (µm)	d_c (µm)	Number of estimates
14	▽	1.1	165	28
	▼	1.1	306	22
	□	1.33	306	6
	■	1.04	532	6
20	◇	0.72	108	6
	◆	0.9	108	6
	○	1.15	108	6
	●	1.10	600	6
	◪	1.35	600	6
	◪	1.75	600	6
43	△	0.8	425	3
	▲	1.6	425	3

Fig.29. Comparison of estimates for collection rate constant, κ, in fluidised beds with values for corresponding fixed beds, after Thambimuthu (ref.14).

confirm that the performance of a fluidised filter at elevated temperature can be predicted from measurements at ambient temperature. Figures 30 to 33 show the results of such calculations for spherical collector of density 2500 kg m^{-3}, filtering particles of density 2500 kg m^{-3} from air. Taking the void fraction at minimum fluidisation, ε_{mf}, as 0.4, the minimum fluidising velocity is calculated from eqns. (27) to (29); resulting values for three levels of temperature and pressure are shown in Fig.30. Values for the single-particle efficiency, calculated from the correlations in Table 1 using these values for U_{mf}, are shown in Figs.31 to 33 in the form of contour plots in the (d_c, d_a) plane.

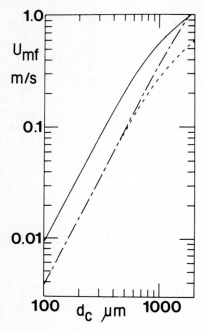

Fig.30. Superficial velocity of air at minimum fluidisation, for spherical collector of density 2500 kg m^{-3}

———————— 300K, 1 bar

—·—·—·— 1100K, 1 bar;

— — — — — 1100K, 10 bar.

The dominant collection process in the bubbling part of the filter is inertia, diffusion, or direct interception depending on the operating conditions and on the characteristics of aerosol and collector, and Figs.31 to 33 show ranges of (d_c,d_a) in which one mechanism contributes more than half the total efficiency. The figures also show the "most penetrating" aerosol diameter; i.e. the value of d_a for which E is a minimum at given d_c. For the reasons discussed in Section II.1, temperature has a strong effect on collection efficiency, whereas the effect of pressure is most marked for collector of order 1mm or more in diameter and results primarily from the effect on U_{mf}.

In general, f_b is affected both by interphase transfer and by collection in the particulate phase. A full analysis of the experimental data to account for these two effects has yet to be carried out. However, Figs.20 to 22 show Ghadiri's calculations (ref.38) for some of the data reported by Doganoglu, with the collection rate in the particulate phase evaluated as for the equivalent fixed bed. The data for 596μm collector (Fig.20) agree well with the predictions of the model in Section III.3 when the crossflow factor, X, is estimated by Darton's approach (ref.37). However, data for 108μm collector (Figs.21 and 22) suggest much larger values for X. Darton's approach allows for the effect of bubble coalescence on mean bubble size, but does not account for the interphase transfer which occurs at coalescence (see Section III.2). Figures 21 and 22 show that reasonable agreement is obtained if the crossflow factor is estimated by

$$X' = X + 20n_c \tag{94}$$

where X is the Darton value and n_c is the number of coalescences experienced by a bubble in its passage through the bed. Eqn.(94) implies an unrealistically large effect of coalescence on interphase transfer. Thus the reason for the discrepancy awaits satisfactory explanation.

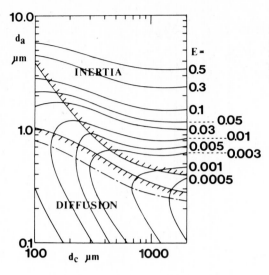

Fig.31. Single-particle collection efficiency in particulate phase of beds of spherical collector: air at 300K and 1 bar; $\rho_a = \rho_c = 2500$ kg m^{-3}.

— ·— ·— · Most penetrating aerosol diameter.

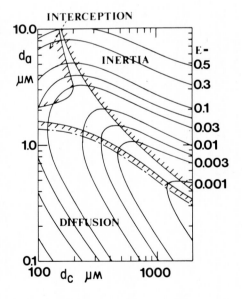

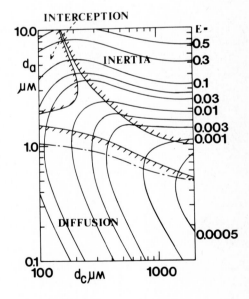

Fig.32. Air at 1100K and 1 bar   Fig.33. Air at 1100K and 10 bar

Single-particle collection efficiency in particulate phase of beds of spherical collector: $\rho_a = \rho_c = 2500$ kg m^{-3}.

— ·— ·— · Most penetrating aerosol diameter.

V.   VARIANTS ON BASIC FLUIDISED FILTER

V.1   Electrofluidised bed

The electrofluidised bed is a device in which electrical forces are utilised to collect dust particles on the granular material forming the fluidised bed. The general idea appears to have been suggested by Johnstone (ref. 55), but the device has been developed further by Melcher and his co-workers (refs. 56-63). An electrofluidised bed dust collector is essentially an electrostatic precipitator with a fluidised bed as the collection zone. The particles to be filtered are first charged by passing the dusty gas through a corona discharge. The gases then pass to the collection bed, which contains an array of electrodes arranged either as vertical plates within or outside the bed ("Crossflow", Fig. 34a) or as horizontal wire grids within the bed ("Coflow", Fig. 34b). An a.c. or d.c. potential is applied between the electrodes, so that dust is collected by attraction between the charge on the particles and the dipoles induced on the collector by the applied field.

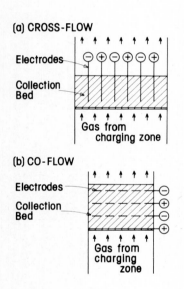

Fig.34 Electrofluidised bed (schematic)

Zahedi and Melcher (refs. 57,58) have given an analysis of filtration in fluidised beds which is analogous to that for conventional fluidised filters given by Doganoglu *et al.* (ref. 36) and extended in Section III above. A simple analysis of flow around a granule in a d.c. field (ref. 57) showed that the single particle collection efficiency resulting from charge-dipole attraction is given approximately by

$$E_E = q_a E'F/\pi\mu d_c U_{mf} \qquad (95)$$

where q_a is the charge on a collected particle and E' is the applied field intensity. As usual for processes occurring in the particulate phase of a bubbling bed, the characteristic velocity in eqn.(95) is the minimum fluidisation velocity, although U_{mf} may be affected by the applied field (refs. 56, 62). The value of E_E can be much larger than the total collection efficiency for all non-electrical processes so that, in terms of the analysis in Section III, the dimensionless filtration rate constant, k, can be much greater than in a conventional fluidised bed. An electrofluidised bed can therefore operate in the range where collection is limited by transfer between the bubbles and the particulate phase (*cf*. Figs. 17 and 18). Zahedi and Melcher (ref. 59) concluded that collection in their

laboratory experiments was in fact limited by interphase transfer, and also found
that horizontal baffles, such as the electrodes in the coflow configuration,
improved filtration by breaking up bubbles.

Alexander and Melcher (ref. 59) extended the measurements and analysis to
applied a.c. fields. A "cut-off" frequency was predicted above which the collection
efficiency declines because the motion of a dust particle is reversed by the
alternating field before the particle can contact the collector:

$$f_c' = \frac{1}{2\pi} \sqrt{\frac{3U_{mf}E'q_a g(\varepsilon)F}{2\mu d_c^3}} \tag{96}$$

where E' is the peak field intensity and $g(\varepsilon)$ is a function of void fraction:

$$g(\varepsilon) = \frac{1 - x^5}{2 - 3x + 3x^5 - 2x^6} \tag{97}$$

where

$$x = (1-\varepsilon_{mf})^{1/3} \tag{98}$$

In addition to enhancing collection, the imposed field can have a very strong
effect on flow in the bed (refs. 56, 58, 61, 62). At low field strengths, the
bed behaves in the conventional way. At higher strengths, of order 10^5 V/m, the
collector granules tend to form "strings" between the electrodes and the bubble
shape is modified. At even higher field strengths, the "strings" become permanent,
so that the particles "flutter" in the gas flow but show no overall bulk movement.
Operation in this regime is undesirable, because collection efficiency is reduced
by gas channelling and by the reduced effective surface area of collector available
for filtration.

Electrofluidised beds are potentially much more compact than conventional
electrostatic precipitators (ref. 60). Moreover, operation with mains frequency
a.c. is possible, because the "cut-off" frequency normally exceeds 50 or 60 Hx
(*cf*. eqn.(96)). Whether an electrofluidised bed offers an advantage over a
conventional fluidised filter appears to depend on process requirements. Zahedi
and Melcher (ref. 58) worked with liquid aerosols in beds up to 0.08 m deep and
with high free-area gas distributors. For 1μm aerosol, the overall penetrations
were comparable to those reported for conventional beds designed to give high
collection in the distributor jets (*cf*. Section IV.2); for sub-micron aerosol,
the electrofluidised bed gave significantly lower penetration. Thus if particles
in the micron range are to be collected and the pressure drop associated with a
low free-area distributor can be tolerated, there seems to be little benefit in
the additional complexity of the electrofluidised bed. However, if pressure drop
is critical or sub-micron particles are to be collected, then an electrofluidised

bed is likely to be preferred. In each case, it is necessary to ensure that collected particles are retained in the bed, and this practical problem is discussed in Section VI.1.

V.2   Rotating fluidised bed

The rotating bed is a device which has been proposed to widen the range of gas velocities which can be handled by a fluidised bed. The bed of granular material forms an annulus contained in a cylindrical vessel rotated about its axis of symmetry. The cylindrical wall forms the gas distributor, and the bed particles are fluidised against their centripetal acceleration by a gas flowing radially inwards. Rotating fluidised beds were first investigated for possible application in hydrogen-cooled nuclear rocket drives (ref. 64). More recently, the technique has been under development for combustion of coal and other fuels (e.g. refs. 65-67). One of the main advantages in this application lies in the potential "turndown" capacity: the speed of rotation can be controlled to permit wide variations in gas flow and fast response to changes in load. Subzwari and Swithenbank (ref. 67) and Chevray *et al*. (ref. 68) have reported studies of the motion of bubbles and entrained particles in rotating fluidised beds.

It should also be possible to use a rotating fluidised bed for gas filtration. Boubel and Junge (ref. 69) showed that the technique is feasible, but their experiments were of limited scope. Some general conclusions can be drawn from the analysis in Sections III and IV, complementing the feasibility study carried out by Pfeffer and Hill (ref. 70). The higher effective "g" values and higher gas velocities, characteristic of rotating beds, both increase the efficiency of inertial collection in the distributor jets (*cf*. eqn.(81) and Fig.26). The value of U_{mf} is increased by the higher "g" (*cf*. eqns. (26) to (30)), and the corresponding effect on collection in the particulate phase differs between the different mechanisms. From the results summarised in Table 1, increasing U_{mf} decreases the efficiency of collection by Brownian diffusion, has little effect on direct interception, gives a weak increase in the efficiency of gravitational settling, and greatly enhances inertial deposition. Increasing U_{mf} also increases the rate of transfer of aerosol between bubbles and particulate phase (see Section III.2). Taken together, these effects show that, for particles larger than typically 1μm in diameter which are collected primarily by inertia (*cf*. Figs.31 to 33), the efficiency of a rotating fluidised bed should be higher than that of a conventional fluidised filter. Smaller particles, collected predominantly by diffusion with efficiency limited by collection rather than interphase transfer, should be filtered less efficiently in a rotating bed. Whether the problem of retention is less serious in a rotating bed remains to be determined.

VI.   APPLICATIONS OF FLUIDISED FILTRATION

VI.1   Gases derived from coal

   Since fluidised bed filters can collect particles a few microns in diameter
with high efficiency, their potential application appears to lie in tertiary
cleaning of gases from which larger particles have already been removed, for
example by passage through one or more stages of cyclones. Compared to conventional equipment for collecting fine particles, fluidised beds are potentially
more compact (see ref. 36) and can employ a wider range of construction materials.
Much current interest therefore lies in their use for cleaning gases derived from
combustion or gasification of coal, to enable the gas either to be passed through
a turbine (ref. 71) or to be vented to atmosphere (ref. 63). For such applications, it is essential to ensure that collected dust particles are retained by
the bed material. Pilney and Erickson (refs. 72,73) investigated the removal of
coal fly ash from air at ambient temperature in fluidised beds, and showed that
capillary forces can be used to enhance retention. At relative humidity greater
than 90%, fly ash was found to agglomerate on to silica sand or pre-formed agglomerates of fly ash. This technique cannot be employed at elevated temperature,
where humidity alone cannot supply the necessary capillary forces. However,
Pilney and Erickson also showed that retention is enhanced by using as collector
either particles which are inherently "sticky" (e.g. wax) or porous particles
which are rendered "sticky" by filling the pores with a non-volatile liquid.
Doganoglu (refs. 20, 36) subsequently showed that non-porous particles could also
be rendered "sticky" conveniently by using them as a fluidised filter to collect a
liquid retention aid dispersed as an aerosol in the fluidising gas. If the
liquid loading on the bed particles was too high fluidisation was lost, but good
retention of solid methylene blue particles on glass ballotini was achieved at
loadings low enough not to impair fluidisation. Ghadiri (ref. 38) investigated
the use of liquid retention aids in beds of silica sand at temperatures around
1100K. The efficiency of retention was found to increase with the loading of
retention aid per unit surface area of collector. Larger collector particles
were found to be preferred, because they can sustain higher surface loading
without losing fluidisation.

   Ciborowski and Zakowski (ref. 49) and Pilney and Erikson (refs. 72,73) found
that retention of dust was enhanced by using collector particles which become
charged by triboelectrification during fluidisation. It is not known whether
such an effect can occur at high temperatures. Melcher et al. (ref. 63) have
proposed the use of a liquid retention aid in an electrofluidised filter,
suggesting that electrical effects alone do not give sufficient retention in
such a device.

Regeneration or disposal of the spent collection medium has received relatively little attention. Ideally the filter could be run in a self-agglomerating mode, to give a disposable granular product, or operated so that the spent collector can be burned or gasified. The configuration suggested by Melcher *et al.* (ref. 63) for cleaning stack gases uses a conventional fluidised bed to remove micron-sized fly ash, followed by an electrofluidised bed to remove submicron ash. The medium in each bed is pulverised coal with an unspecified oil as retention aid, so that the spent collector is added to the combustor feed. Unfortunately this ingenious solution cannot be applied at temperatures where the coal will burn, devolatilise, or gasify; an inert collector must then be used.

VI.2   Other applications

Two applications of fluidised filtration are known to have been operated successfully in processes other than cleaning gases from coal. It is interesting to note that, in each case, a shallow fluidised bed was used for reasons associated with other process requirements, and subsequently proved to be effective for particulate collection.

The "ALCOA 398" process (ref. 74), used to recover fluoride from gases from aluminium smelting, is shown schematically in Fig.35. It contains a shallow fluidised bed of metal-grade alumina, followed by a conventional bag filter. Dust collected by the bags is returned to the fluidised bed. The alumina bed was originally intended to remove hydrogen fluoride from the waste air stream, and pilot studies indicated that its capacity was 1.8% HF by weight. Particulate fluorides were to be collected by the bag filters.

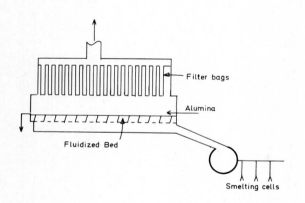

Fig.35 "ALCOA 398" process (schematic)

However, in operation the bed particles were found to collect up to 3.5% fluoride as HF, while the load on the bag filters was much less than expected. Thus primary interception of particulate fluorides must occur in the bed, acting as a fluidised filter. It is also interesting to note that the temperature of operation is limited to below 135°C by the bag material, and that most of the maintenance cost is associated with the bag filter. Alumina is fed continuously to the bed, and spent alumina passes directly for smelting so that the recovered fluoride is recycled without regeneration or disposal problems. It is reported that the savings in fluoride more than offset the operating cost of the process.

This approach was developed further as the "ALCOA 446" process (ref. 75) for collecting fumes from production of prebaked anodes for aluminium reduction cells. The fumes contain hydrocarbon vapours, some coke dust, and lesser quantities of fluorides. Like the 398 process in Fig.35, a shallow fluidised bed of alumina is used followed by a bag filter. In addition to fluoride collection as in the 398 process, the alumina now serves to collect the hydrocarbons and coke. Regeneration is achieved by incinerating the bed material. By maintaining a high hydrocarbon content on the alumina, the incinerator is made self-sustaining. Regenerated alumina is either recycled through the collection process, or passed forward for smelting.

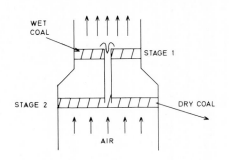

Fig.36 Two-stage fluidised bed drying of coal (schematic)

Fluidised filtration has been employed fortuitously in the U.S.S.R. for coal drying. Rubin and Margolin (ref. 76) describe a two-stage dryer, shown schematically in Fig.36, designed to reduce dust emissions from dry coal by operating the lower "dry" stage at reduced gas velocity. In operation, dust emission from the two-stage dryer proved to be lower than from a single-stage dryer followed by two stages of cyclones. Evidently dust entrained from the lower stage was collected in the upper "wet" stage, with the 11 to 15% moisture content acting as a natural retention aid.

An unsuccessful trial reported by Svrcek and Beeckmans (ref. 77) demonstrates the importance of ensuring retention of collected particulates. A four-stage filter of fluidised sand was used to collect particles from Kraft furnace effluent gas. The results are difficult to interpret, since the size of collected particles was unknown, but performance appeared to be limited by re-entrainment: the overall collection efficiency ranged from 24 to 81%, and declined with increasing gas velocity. Svrcek and Beeckmans concluded that a fluidised bed is less effective than a wet scrubber for this application.

It is also possible to use a fluidised bed as an agglomerator rather than a filter. With operation in this mode, there is no replacement of the bed material. Fine dust particles are collected in the usual way and retained in the bed long enough to agglomerate with other collected particles, so that on re-entrainment the particles are sufficiently large to be removed from the gas by a device such as a cyclone. In Britain, this approach is being investigated at Warren Spring Laboratory for suppression of fume from metals extraction (ref. 78). In Italy, work at the Politecnica di Napoli is directed at removal of particulates from diesel exhaust (ref. 79).

NOTATION

A	Cross-sectional area of filter
C	Aerosol concentration, expressed as number of particles per unit gas volume
C_H	Average aerosol concentration in gas leaving fluidised bed
D	Brownian or molecular diffusivity
d_a	Diameter of aerosol particle
d_B	Volume-equivalent bubble diameter
d_c	Diameter of collector particle
E	Single particle collection efficiency, defined by eqn.(11)
E'	Intensity of applied electric field
F	Stokes-Cunningham slip correction factor
f	Overall penetration through filter, defined by eqn.(17)
f_b	Penetration through bubbling part of fluidised bed, defined by eqn.(63)
f_j	Penetration of aerosol through grid jets, defined by eqn.(53)
f'_c	"Cut-off" frequency, above which collection efficiency caused by a.c. electric field declines
Ga	Galileo number, defined by eqn.(28)
g	Gravitational acceleration
H	Total filter depth
h	Distance from filter inlet
J	Total net rate of transfer of aerosol from bubble to particulate phase
K	Interphase transfer coefficient per unit volume of bubble phase
k	Dimensionless filtration rate constant, defined by eqn.(56)
k_B	Coefficient of diffusional transfer between bubble and particulate phase in fluidised bed; Boltzmann's constant (eqns.(1) and (2))
k_D	Coefficient of mass transfer to surface of collector particle
M	Molecular weight of gas
m_1, m_2	Roots of eqn.(62)
N_{DI}	Interception parameter, defined by eqn.(10)
N_G	Gravitational settling parameter, defined by eqn.(7)
N_{or}	Dimensionless group defined by eqn.(81)
N_{uc}	Dimensionless terminal velocity of collector particle, defined by eqn.(83)
n	Number of aerosol particles per unit time presented to single layer of collector particles
n_c	Number of coalescences experienced by a bubble on passage through bed
P	Gas pressure
q	Volumetric gas flowrate through bubble
q_a	Electrical charge on collected dust particle
R	Number of aerosol particles per unit time collected in unit volume of fixed bed or particulate phase; universal gas constant (eqn.(2))
Re_c	Collector Reynolds number, defined by eqn.(16)

r   Number of aerosol particles per unit time collected by individual collector particle

S   Surface area of bubble

Sc   Schmidt number, defined by eqn.(14)

St   Stokes number, defined by eqn.(6)

St_{or} Stokes number in distributor orifice, defined by eqn.(82)

T   Gas temperature

U   Superficial velocity of gas through filter

u_t Terminal velocity of aerosol particle in gas

u_{tc} Terminal velocity of collector particle in gas

V   Bubble volume

X   Crossflow factor, defined by eqn.(39)

x $(1-\varepsilon_{mf})^{1/3}$

Y   Dimensionless group describing transfer from bubble phase to single layer of collector particles, defined by eqn.(42)

y   Distance travelled by bubble through fluidised bed

z   Dimensionless distance from end of grid jets, defined by eqn.(55)

α   Dimensionless group describing filter depth associated with single layer of collector particles, defined by eqn.(41)

β   Fraction of fluidising gas passing through bed in bubble phase (eqn.(31))

γ   Number of layers of collector particles in depth $(H-h_j)$

δ   Fraction of aerosol collected on entry to or exit from fixed bed

ε   Void fraction in filter medium

ε_B Time-averaged fraction of cross-sectional area of fluidised bed occupied by bubble phase

$\bar{\varepsilon}_B$ Average value of ε_B over whole depth of bed

η   Overall filter efficiency, (1-f)

κ   Rate constant for filtration, defined by eqn.(22)

λ   Molecular mean free path

μ   Gas viscosity

ρ_a Density of aerosol particle

ρ_c Density of bed particle

ρ_g Density of gas

φ   Fraction of original bubble phase gas remaining within that phase

Suffices

B	Bubble phase in fluidised bed	mf	Minimum fluidisation conditions
D	Diffusion	o	Entry to filter bed
DI	Direct interception	P	Particulate phase in fluidised bed
E	Charge-dipole attraction	1	Exit from fixed bed filter.
G	Gravitational settling		
I	Inertial impaction		
j	Conditions at end of gas jets		

122

REFERENCES

1   H.P.Meissner and M.S.Mickley, Ind.and Eng.Chem., 41(1949)1238-1242.
2   C.N.Davies (Ed.), Aerosol Science, Academic Press, New York, 1966.
3   R.Clift, J.R.Grace and M.E.Weber, Bubbles Drops and Particles, Academic Press, New York, 1978.
4   C.N.Davies, Proc.Phys.Soc.London, 57(1945)259-270.
5   B.E.Dahneke, J.Aerosol Sci., 4(1973)139-145.
6   K.V.Beard, J.Atmos.Sci., 33(1976)851-864.
7   K.V.Thambimuthu, B.K.C.Tan and R.Clift, Instn Chem.Engrs Symp.Ser. no.59, 1980, paper 1:3.
8   A.Meisen, A.J.Bobkowicz, N.E.Cooke and E.J.Farkas, Can.J.Chem.Eng., 49(1971)449-457.
9   L.C.Paretsky, L.Theodore, R.Pfeffer and A.M.Squires, Jl.Air Pollution Control Assoc., 21(1971)204-209.
10 L.C.Paretsky, Ph.D. Thesis, The City University of New York (1972).
11 K.V.Thambimuthu, Y.Doganoglu, T.Farrokhalaee and R.Clift, in Symp.on Deposition and Filtration of Particles from Gases and Liquids, Loughborough, Sept.6-8, 1978, Society of Chemical Industry, pp 107-119.
12 R.G.Dorman, in E.G.Richardson (Ed.), Aerodynamic Capture of Particles, Pergamon, London, 1960, pp 112-122.
13 N.A.Fuchs, The Mechanics of Aerosols, Pergamon, London, 1964.
14 K.V.Thambimuthu, Ph.D. Dissertation, University of Cambridge, 1980.
15 R.Rajagopalan and C.Tien, in R.J.Wakeman (Ed.), Progress in Filtration and Separation, Vol.1, Elsevier, Amsterdam, 1979, pp 179-269.
16 E.J.Wilson and C.J.Geankoplis, Ind.and Eng.Chem.Fundamentals, 5(1966)9-14.
17 J.Gebhart, C.Roth and W.Stahlhofen, J.Aerosol Sci., 4(1973)355-371.
18 M.Balasubramanian and A.Meisen, J.Aerosol Sci., 6(1975)461-463.
19 C.N.Davies, Air Filtration, Academic, London, 1973.
20 Y.Doganoglu, Ph.D. Thesis, McGill University, 1975.
21 J.F.Davidson and D.Harrison, Fluidised Particles, Cambridge University Press, 1963.
22 J.F.Davidson and D.Harrison (Eds), Fluidization, Academic Press, New York, 1971.
23 D.Kunii and O.Levenspiel, Fluidization Engineering, Wiley, New York, 1969.
24 S.Ergun, Chem.Eng.Prog., 48(1952)89-94.
25 C.Y.Wen and Y.H.Yu, AIChEJ, 12(1966)610-612.
26 D.Geldart, Powder Technology, 7(1973)285-292.
27 R.D.Toomey and H.F.Johnstone, Chem.Eng.Prog., 48(1952)220-226.
28 J.F.Davidson and D.Harrison, Chem.Eng.Sci., 21(1966)731-737.
29 J.R.Grace and R.Clift, Chem.Eng.Sci., 29(1974)327-334.
30 M.J.Lockett, J.F.Davidson and D.Harrison, Chem.Eng.Sci., 22(1967)1059-1066.
31 R.Clift, J.R.Grace, L.Cheung and T.H.Do, J.Fluid.Mech., 51(1972)187-205.
32 J.C.Orcutt, J.F.Davidson and R.L.Pigford, Chem.Eng.Prog.Symp.Ser. no.38, 58(1962)1-15.
33 J.R.Grace, Chem.Eng.Sci., 26(1971)1955-1957.
34 C.Chavarie and J.R.Grace, Ind.and Eng.Chem.Fundamentals, 14(1975)75-91.
35 J.M.D.Merry, AIChEJ, 21(1975)507-510.
36 Y.Doganoglu, V.Jog, K.V.Thambimuthu and R.Clift, Trans.Instn Chem.Engrs, 56(1978)239-248.
37 R.C.Darton, Trans.Instn Chem.Engrs, 57(1979)134-138.
38 M.Ghadiri, Ph.D. Dissertation, University of Cambridge, 1980.
39 D.S.Scott and D.A.Guthrie, Can.J.Chem.Eng., 37(1959)200-203.
40 C.H.Black and R.W.Boubel, Ind.and Eng.Chem.Proc.Des.and Dev., 8(1969)573-578.
41 W.Jugel, E.D.Reher, E.Grobler and A.Tittman, Chem.Tech., 22(1970)403.
42 G.Dumont, W.Goossens, A.Taeymans and W.Balleux, Het Ingenieursblad, 10(1973) no.3, 3-7.
43 P.Knettig and J.M.Beeckmans, Can.J.Chem.Eng., 52(1974)703-706.
44 M.L.Jackson, AIChE Symp.Ser. no.141, 70(1974)82-87.
45 D.McCarthy, A.J.Yankel, R.G.Patterson and M.L.Jackson, Ind.and Eng.Chem. Proc.Des.and Dev., 15(1976)266-272.

46 R.G.Patterson and M.L.Jackson, AIChE Symp.Ser. no.161, 73(1977)64-73.
47 A.R.Figueroa and W.Licht, paper 33b, 81st Natl AIChE Meeting, 1976.
48 R.K.Chaturvedi and J.C.Reed, Proc. 2nd Pacific Chem.Eng.Congress, 1977.
49 J.Ciborowski and L.Zakowski, International Chem.Eng., 17(1977)529-548.
50 B.K.C.Tan, Personal Communication, 1980.
51 P.N.Rowe, H.J.MacGillivray and D.J.Cheesman, Trans.Instn Chem.Engrs, 57(1979)194-199.
52 M.Ghadiri and R.Clift, Ind.and Eng.Chem.Fundamentals, 19(1980)440.
53 S.Donadono and L.Massimilla, in J.F.Davidson and D.L.Keairns (Ed.), Fluidization, Cambridge University Press, 1978, pp 375-380.
54 G.Donsì, L.Massimilla and L.Colantuoni, in J.R.Grace and J.M.Matsen (Ed.), Fluidization, Plenum, New York, 1980, pp 297-304.
55 H.F.Johnstone, U.S.Patent 2,924,294 (1960).
56 T.W.Johnson and J.R.Melcher, Ind.and Eng.Chem.Fundamentals, 14(1975)146-153.
57 K.Zahedi and J.R.Melcher, Jl Air Pollution Control Assoc., 26(1976)345-352.
58 K.Zahedi and J.R.Melcher, Ind.and Eng.Chem.Fundamentals, 16(1977)248-254.
59 J.C.Alexander and J.R.Melcher, Ind.and Eng.Chem.Fundamentals, 16(1977)311-317.
60 J.R.Melcher, K.S.Sachar and E.P.Warren, Proc.IEEE, 65(1977)1659-1669.
61 P.W.Dietz and J.R.Melcher, AIChE Symp.Ser. no.175, 74(1978)166-174.
62 P.W.Dietz and J.R.Melcher, Ind.and Eng.Chem.Fundamentals, 17(1978)28-32.
63 J.R.Melcher, J.C.Alexander and K.Zahedi, U.S.Patent 4,154,585 (1979).
64 L.P.Hatch, W.H.Regan and J.R.Powell, Nucleonics, 18(1960) no.12, 102-3.
65 N.Demircan, B.M.Gibbs, J.Swithenbank and D.S.Taylor, in J.F.Davidson and D.L.Keairns (Ed.), Fluidization, Cambridge University Press, 1978, pp 270-275.
66 C.I.Metcalfe and J.R.Howard, in J.F.Davidson and D.L.Keairns (Ed.), Fluidization, Cambridge University Press, 1978, pp 276-279.
67 M.P.Subzwari and J.Swithenbank, Inst.Energy Symp.Ser. no.4, 1980, paper VI-3.
68 R.Chevray, Y.N.I.Chan and F.B.Hill, AIChEJ, 26(1980)390-398.
69 R.W.Boubel and D.C.Junge, AIChE Symp.Ser. no.128, 69(1973)138-141.
70 R.Pfeffer and F.B.Hill, Brookhaven National Lab.Rept, no.BNL 50990 (1978).
71 R.Clift, M.Ghadiri and M.J.Cooke, U.K.Patent Application GB 2,014,472A (1979).
72 J.P.Pilney and E.E.Erickson, Jl Air Pollution Control Assoc., 18(1968)684-685.
73 J.P.Pilney and E.E.Erickson, U.S.Dept of Interior, Bureau of Mines, final report on contract 14-69-0070-375 (1968).
74 C.C.Cook, G.R.Swany and J.W.Colpitts, Jl Air Pollution Control Assoc., 21(1971)479-483.
75 M.W.Wei, AIME Meeting on "Light Metals", 1975, Vol.2, pp 261-278.
76 Yu.M.Rubin and Yu.A.Margolin, Coke and Chem. (USSR), 17(1974) no.3, 15-17.
77 W.Y.Svrcek and J.M.Beeckmans, Trans.Am.Pulp and Paper Inst. 59(1976) no.9, 79-82.
78 P.R.Dawson, Personal Communication, 1980.
79 L.Massimilla, Personal Communication, 1980.

SEPARATION OF PARTICULATE MIXTURES IN LIQUID DIELECTRIC AND MAGNETIC MEDIA

U.Ts. ANDRES

Department of Mineral Resources Engineering, Imperial College of Science and Technology, London SW7 2BP, England.

CONTENTS

ABSTRACT

The orientation, spatial displacement and interaction of solid particles, in vacua and in liquid media, polarised in electrical and in magnetic fields, are computed. The separation effect in mixtures of particulate solids in the super-imposed electromagnetic and gravitational fields is then analysed and the simple graphico-analytical method of positioning particles in superimposed fields is developed, with examples of such separations.

INTRODUCTION

Process of separation of individual components of mixture in order to obtain elementary materials for a direct use or for manufacturing different composites is one of the most fundamental technological processes and is applicable to mixture of substances in gaseous, liquid and solid states as well as to their various combinations.

While the separation of gases or liquids is carried out on the basis of atomic (isotopic) and molecular identity, the separation of solids is carried out on the basis of the predominance in the content of identical molecules or crystalines in solid particles and realized in the form of its spatial grouping.

The use of electrical and magnetic fields for the spatial grouping solid particles in liquid polarisable media provides special physical conditions for simultaneous separation of unlimited number of components according to their permittivity, magnetic susceptibility and density. These separation processes, being dielectric medium separation (DMS) and magnetohydrostatic separation (MHS) are based on selective polarisation capacity of different solid components of mixtures, in electrical and magnetic fields respectively and have similar physical principles carried out in resembling separating devices.

The conceptual design of the displacing particles' trajectories in both continuous separation processes is based on interaction of ponderomotive forces with gravitational, frictional and drag forces affecting particles on their way towards separate exits from the separation space.

It is essential that both processes have a two-product-separation version
where the split of components is carried out on the basis of difference in the
permittivity or magnetic susceptibility of particles disregarding their density
as a parameter of separation. So, both processes can be used in extraterristrial
conditions of low or zero gravitation.

The prevalent particles' trajectories are determined by design of separation
dynamics for individual particles, but the real behaviour of particulate solids
at concentration of the order of 10^6-10^8 fragments per cm^3 includes incongruent
motions of solids depending on stochastic nature of inertial and electromagnetic
interraction among particles. In order to decrease the erroneous displacement
of particles in DMS and MHS separation usual measures of increase of energy
level of interaction between electromagnetic and gravitational forces, ensuring
the undisturbed flow of particles as well as repetitions of separation act for
primary products are necessary.

We proceed from the assumption that the separation displacement is a result
of orientational-translational motions and attachments of solid particles due to
their interaction with external polarizing fields and their dipole-dipole interactions. So, in this paper we will discuss only the main dynamic phenomena
producing the spatial separation of solid particles resulting from the propagation
of electrical and magnetic fields in the system consisting of a polarized
particulate mixture immersed in a polarized liquid medium. Each elementary act
of spatial separation of solid particles will be discussed.

1. ROTATIONAL ORIENTATION OF SOLID PARTICLES
a) Rotation of solid particle in the magnetic field.

Orientational rotation of anisotropic fragments is the more general of the
two kinds of particle motions (the other being translation) taking place in
magnetic and electrical fields, both uniform and non-uniform. If magnetic moment
$\bar{m}$ is induced by an external magnetic field $\bar{H}_e$ (where $\bar{H}_e = 0$, $\bar{m} = 0$), $\bar{m}\|\bar{H}_e$ takes
place when one of the three axes of orientated particle is parallel to $\bar{H}_e$.

Let us consider a solid elliptical particle of flattened tablet shape, with
two axes longer than the third, and exposed to orientation in a magnetic field.
Although there are three possible orientations for a triaxial ellipsoid, only
two possibilities are really under question: case a, when the longest axis is
parellel to the field, and case b, when the shortest axis coincides with the
direction of $\bar{H}_e$ (see Fig.1.). The equilibrium orientation of the ellipsoid can
be found from estimation of the direction of the minimum energy requirement.

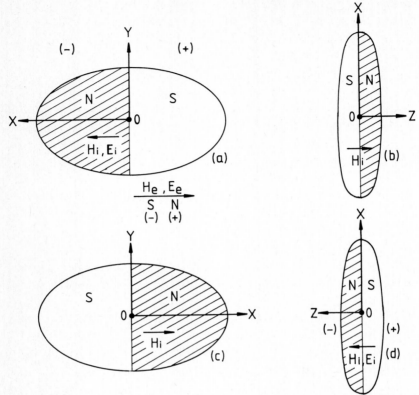

Fig. 1.   Orientation of solid anisometric particles in magnetic and electrical fields.

The energy of a particle with a magnetic moment $\bar{m}$, in an external field $\bar{H}_e$, is their scalar product:

$$W = - \bar{m} \cdot \bar{H}_e = - m \cdot H \cos \alpha \qquad (1)$$

The moment $\bar{m}$ is the magnetisation M of a particle of volume V.

When $\alpha = 0$ and $\cos \alpha = 1$, the positional energy reaches its minimum. Therefore

$$\bar{m} = \bar{M}V \qquad (2)$$

and the magnetisation of a unit volume of particle is

$$\bar{M} = \chi \bar{H}_i \qquad (3)$$

(Here χ is the magnetic susceptibility of a particle, and $\bar{H}_i$ is the magnetic field inside the particle.)

This internal magnetic field can be found from the boundary conditions of the Maxwell equations. So, from

$$\text{rot } \bar{H} = \frac{4\pi}{c} \bar{j} \tag{4}$$

$$\text{div } \bar{B} = 0$$

it follows respectively that

$$\bar{H}_e^t = \bar{H}_i^t$$

$$\bar{B}_e^n = \bar{B}_i^n \tag{5}$$

where the superscripts "t" and "n" correspond respectively to the tangential and normal components of the fields, and of the induction, which appear at the boundaries between solid particles and surrounding continuous medium.

Therefore for orientation (a) in Fig. 1,

$$H_e^t = H_i^t \text{ and } M = \chi H_i = \chi H_e$$

The magnetic moment m is given by

$$m = \chi H_e V \tag{6}$$

The energy position is

$$W_a = -(\chi H_e V) H_e \doteq -\chi V H_e^2 \tag{7}$$

For orientation (b) in Fig. 1, we have

$$B_e^n = B_i^n$$

In vacuum $B_e^n = H_e$ and $B_i = \mu H_i$. Then,

$$H_e = \mu H_i \tag{8}$$

and

$$H_i = H_e/\mu$$

Here μ is the magnetic permeability of the particle.

Finally

Finall

$$m = \chi H_i V = \chi \frac{H_e}{\mu} V \tag{9}$$

and the energy is then

$$W_b = -\{\chi \frac{H_e}{\mu} V\} H_e = \chi \frac{V H_e^2}{\mu} \tag{10}$$

Substituting (7) into (1), we have

$$W_b = -\frac{1}{\mu} W_a \tag{11}$$

For a paramagnetic particle $\{\chi > 0\}$, the permeability is given by $\mu = 1 + 4\pi\chi > 0$ and the absolute value of the energy, is

$$|W_b| \mu = |W_a|. \tag{12}$$

Then $|W_a| > |W_b|$ and $W_a < W_b$. This means that for a paramagnetic particle, in air or in a vacuum, orientation (a) is energetically more advantageous.

For a diamagnetic particle $\chi < 0$ and $\mu = 1 - 4\pi\chi < 0$. Then from (8) it follows that $|W_a| < |W_b|$ and $W_a > W_b$. This means that for a diamagnetic particle in a vacuum or in air, orientation (b) will occur.

As was implied earlier, the above two descriptions of the orientations are true for media with $\chi \simeq 0$, that is vacua or air. If the separation process is carried out in any liquid, then the equilibrium orientation, corresponding to the minimum of positional energy, is a result of a dynamic reciprocity between particle and medium. Then for the orientation of particles in the liquid medium, eqns. (7) and (10) become respectively

$$W_a = -(\chi_p - \chi_\ell) V H_e^2 \tag{13}$$

and

$$W_b = -\frac{(\chi_p - \chi_\ell) V H^2}{\mu_p / \mu_\ell} = \frac{(\chi_p - \chi_\ell) V H_e^2}{(1 + 4\pi\chi_p)(1 + 4\pi\chi_\ell)} - \frac{(\chi_p - \chi_\ell) V H_e^2}{1 + 4 (\chi_p - \chi_\ell)} \tag{14}$$

Here subscripts 'p' and 'ℓ' correspond respectively to particle and liquid.

A paramagnetic ellipsoid in a paramagnetic liquid takes orientation 1a (Fig.1) if $\chi_p - \chi_\ell > 0$ and orientation 1d if $\chi_p - \chi_\ell < 0$. (In diamagnetic liquids, containing paramagnetic particles, the difference in susceptibilities is always positive and orientation 1a occurs invariably.)

A diamagnetic ellipsoid in any paramagnetic liquid is always orientated as shown in 1b. For a diamagnetic ellipsoid in a diamagnetic liquid, there are two possible equilibrium orientations: 1b, for a case where $\chi_p - \chi_\ell < 0$ and 1c, for a case where $\chi_p - \chi_\ell > 0$.

b) Rotation of solid particles in the electrical field

There is a good analogy between the behaviour of polarised dielectrics and magnetics (solid as well as liquid) in the corresponding polarising fields. However, solid and liquid dielectrics are polarised only in the direction opposing that of external polarising field.

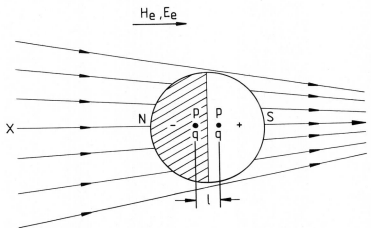

Fig. 2. Spherical particle in the non-uniform polarising fields.

For dielectric particles this results in only two orientations: 1a, where the difference in dielectric constants of particles (ε_p) and liquid (ε_ℓ) is greater than zero, and 1d, where $\varepsilon_p - \varepsilon_\ell < 0$

The orientational motion itself does not produce a separation effect but is a preliminary stage in the dipole-dipole interaction producing the selective clustering. The selective clustering is a basical separational phenomenon in some versions of magnetic and dielectric separation.

2. TRANSLATIONAL FORCES ACTING ON SOLID PARTICLES

The distortion of the uniformity in the magnetic and electrical fields gives rise to net translational forces on the solid particles and liquids placed in these fields, in addition to the orientational revolution of particles in a purely uniform fields. The non-uniformity of polarising fields is a result of the configuration of the electrodes or pole pieces on the one hand and the presence of solid particles with a permittivity or a magnetic susceptibility which differs from that of the liquid media.   The translational forces cause movement of anisometric, as well as isometric particles.   In a vacuum, the rise of magnetic and electrical translational forces acting on solids is a result of the difference between Coulombic forces on the opposite poles of the particles, orientated along the direction of the corresponding non-uniform fields (see Fig. 2).   Thus for a spherical paramagnetic particle of radius R along axis X, there is a rise of the net translational force, which (in c.g.s. units) can be calculated from the difference in Coulombic forces, mentioned above.

Therefore

$$F_x = - pH + p(H + \ell\frac{dH}{dx}) \tag{15}$$

but $p\ell = m$.   Then,

$$F_x = m\frac{dH}{dx} = MV\frac{dH}{dx} = \chi VH = \frac{4}{3}\pi r^3\chi H\frac{dH}{dx} = \frac{2}{3}\pi r^3\chi \text{ grad } H^2 \tag{16}$$

(Here p is the pole strength of a magnetised particle.)

In the liquid magnetic medium the net translational force is a result of the complex Coulombic interaction forces between particles and liquid, and the analogous calculation gives

$$F_x = \frac{2}{3}\pi r^3 (\chi_\ell - \chi_p) \text{ grad } H^2 \tag{17}$$

Here, translation of a particle along the south direction of the polarising external field (see Fig. 2) is accompanied by the reciprocal flow of an equal volume of liquid in the north direction.

The calculation of the difference in the Coulombic forces, acting on opposite poles of a dielectric sphere, immersed in a liquid dielectric polarised by non-uniform electrical field (Fig. 2) gives correspondingly

$$F_x = -qE + q (E + \ell\frac{dE}{dx}) = q\ell\frac{dE}{dx} = p\frac{dE}{dx} \tag{18}$$

(Here q is a charge of particle, E is electrical field and p is a dipole moment of particle.)

Now, the dipole moment of a dielectric sphere is given by

$$p = V\varepsilon_o\varepsilon_\ell E \{ \frac{3(\varepsilon_p - \varepsilon_\ell)}{(\varepsilon_p + 2\varepsilon_\ell)} \} = \frac{4}{3}\pi r^3\varepsilon_o\varepsilon_\ell \{ \frac{3(\varepsilon_p - \varepsilon_\ell)}{(\varepsilon_p + 2\varepsilon_\ell)} \} \tag{19}$$

So the force along the electrical field E is

$$F_x = \frac{4}{3}\pi\varepsilon_o \frac{\varepsilon_\ell(\varepsilon_p - \varepsilon_\ell)}{(\varepsilon_p + 2\varepsilon_\ell)} \; r^3 \; \frac{\text{grad } E^2}{2} \tag{20}$$

The structure $\dfrac{3(\varepsilon_p - \varepsilon_\ell)}{(\varepsilon_p + 2\varepsilon_\ell)}$ in the equation (19) for cylindrical particles should be changed for $2(\varepsilon_p - \varepsilon_\ell)/(\varepsilon_p + \varepsilon_\ell)$ and for the flat one for $(\varepsilon_p - \varepsilon_\ell)/\varepsilon_p$.

3. FORCES IN DIFFERENT FIELD CONFIGURATIONS

a) Forces in the wedge-shaped polarising fields.

Let us consider the forces acting in a wedge-shaped field generated by two plane electrodes mutually inclined at an angle α (Fig. 3).

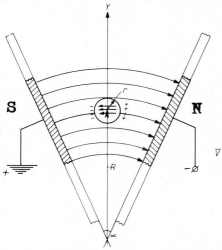

Fig. 3. Spherical particle in the wedge-shaped polarising fields.

It should be noted that the perfect circularity of the force lines is preserved only in the central zone of the wedge-shaped capacitor or air gap of magnetic pole pieces, while at the top and bottom they are distorted by boundary effects. In the central zone along the y-axis, the field is a linear function of the radius R. For the electrical field we have

$$E = \frac{180V}{\pi\alpha} \cdot \frac{1}{R} = \frac{A}{R} \tag{21}$$

where V is the field potential. The field gradient is

$$\text{grad } |\overline{E}| = -\frac{A}{R^2}\overline{j} \tag{22}$$

$\overline{j}$ being the unit field vector directed upward along R. For $r \ll R$, the force acting on a spherical particle is obtained by substituting (21) and (22) in the expression for the ponderomotive force in (20); we have

$$\bar{F} = 4\pi\varepsilon_0\varepsilon_\ell \left(\frac{\varepsilon_p - \varepsilon_\ell}{C_p + 2\varepsilon_\ell}\right) r^3 \frac{A}{R} \left(-\frac{A}{R^2}\right) \bar{j} = -4\pi\varepsilon_0\varepsilon_\ell \left(\frac{\varepsilon_p - \varepsilon_\ell}{\varepsilon_p + 2\varepsilon_\ell}\right) A\frac{2r^3}{R^3} \bar{j} \tag{23}$$

or

$$\bar{F} = -4\pi\varepsilon_0\varepsilon_\ell \left(\frac{\varepsilon_p - \varepsilon_\ell}{\varepsilon_p + 2\varepsilon_\ell}\right) \left(\frac{180}{\pi\alpha}\right)^2 v^2 \frac{r^3}{R^3} \bar{j} \tag{24}$$

The force per unit volume of the sphere is

$$f = \left| 3\varepsilon_0\varepsilon_\ell \left(\frac{\varepsilon_p - \varepsilon_\ell}{C_p + 2\varepsilon_\ell}\right) \left(\frac{180}{\pi\alpha}\right)^2 \frac{v^2}{R^3} \right| \tag{25}$$

or, in CGS units,

$$f = \left| \frac{\varepsilon_\ell}{12\pi} \left(\frac{\varepsilon_p - \varepsilon_\ell}{\varepsilon_p + 2\varepsilon_\ell}\right) \left(\frac{180}{\pi\alpha}\right)^2 \frac{v^2}{10^4 R^3} \right| \tag{26}$$

The wedge-shaped field is suitable for the separation of mixtures containing several components of high density and permittivity, and several others with low ones.

b) Force of quadrupole capacitors

At present, a single pair of poles is used for generating the working field in both electric and magnetic separators. The use of multipole systems in magnetic separators is hampered by design difficulties; on the other hand, electric separators have distinct possibilities in this respect.

Application of fields generated by multipole pairs of electrodes, or by an array of electrodes laid out at different angles relative to each other, is still a matter of theoretical analysis. Suffice it to note here that the displacing force generated by a quadrupole capacitor is twice that of a dipole capacitor with the same field and field gradient. Figure 4 shows, schematically, a quadrupole capacitor consisting of two pairs of inclined electrodes, so that the field space has the form of a truncated pyramid. (When a liquid dielectric of variable permittivity is placed, a uniform prismatic field may replace the non-uniform pyramidal one.) From (20) the force exerted on a sphere by a non-uniform field in a liquid dielectric of homogeneous permittivity is given by:

$$F = 2\pi\varepsilon_0\varepsilon_\ell \frac{\varepsilon_p - \varepsilon_\ell}{\varepsilon_p + 2\varepsilon_\ell} r^3 \operatorname{grad} E^2 \tag{27}$$

or, writing

$$2\pi\varepsilon_o\varepsilon_\ell\frac{\varepsilon_p - \varepsilon_\ell}{\varepsilon_p + 2\varepsilon_\ell} \; r^3 = S \qquad (28)$$

by

$$F = S \; \text{grad} \; E^2 \qquad (29)$$

For the joint action of two equal fields at right-angles, we have

$$E = E' + E'' = \sqrt{2E}' \qquad (30)$$

whence

$$F = S \; \text{grad} \; (\sqrt{2E}')^2 = 2S \; \text{grad} \; (E')^2 \qquad (31)$$

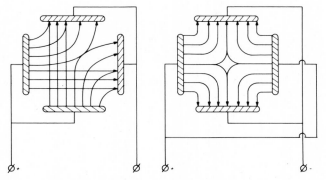

Fig. 4. Electric field of quadrupole capacitor

In a quadrupole capacitor, the resultant field vector is inclined at 45° to the component vectors but this is irrelevant to the direction of the pondero-motive force, which always coincides with that of the field gradient.

For a wedge-shaped quadrupole field, the ponderomotive force is given by

$$\overline{F} = 4\pi\varepsilon_o\varepsilon_\ell \; (\frac{\varepsilon_p - \varepsilon_\ell}{\varepsilon_p + 2\varepsilon_\ell}) \; (\frac{180}{\pi\alpha})^2 \; v^2\frac{r^3}{R^3} \; \overline{j} \qquad (32)$$

c) Translational force in wedge-shaped magnetic field.

The expression for translational force, acting on the solid spherical particle of magnetic susceptibility χ_p in magnetic field H filled with liquid magnetic medium of susceptibility χ_ℓ can be found from similar considerations.

The expression of the force is

$$F = \frac{4}{3} \pi (\chi_\ell - \chi_p) A^2 (r/R)^3 \tag{33}$$

Here, A is the constant of wedge-shaped field which is

$$A = RH = const \tag{34}$$

d) Translational forces in hyperbolic capacitor

In a capacitor formed by the two branches of a symmetric hyperbola (Fig. 5) the field along the y-axis is

$$E = BR \tag{35}$$

and its gradient:

$$grad\ E = B\vec{k} \tag{36}$$

where R is the distance between the sphere centre and the origin, B is the field constant, and $\vec{k}$ is the unit field vector in the y-direction.

The force acting on a spherical particle is

$$\overline{F} = pB\overline{k} \tag{37}$$

or, for a sphere of radius r,

$$F = 4\pi\varepsilon_o\varepsilon_\ell \left(\frac{\varepsilon_p - \varepsilon_\ell}{\varepsilon_p + 2\varepsilon_\ell}\right) r^3 BR.B\vec{k} \tag{38}$$

In order to determine the constant B, we assign the y-axis to the line of zero potential ($\psi = 0$). In that case all hyperbolas xy = const. lie on equipotential surfaces; indeed,

$$\psi = Bxy \tag{39}$$

satisfied the Laplace equation:

$$\nabla^2\psi \left(\frac{\partial}{\partial x^2} + \frac{\partial}{\partial y^2}\right) Bxy = 0 \tag{40}$$

The constant B depends on the configuration of the electrode and on the applied voltage V.

At the positive electrode,

$$\psi = \frac{V}{2} = Bxy = B\frac{a^2}{2} \tag{41}$$

Hence

$$B = \frac{V}{a^2} \tag{42}$$

which, substituted in (17), yields the force along the y-axis:

$$F = 4\pi\varepsilon_o\varepsilon_\ell \left(\frac{\varepsilon_p - \varepsilon_\ell}{\varepsilon_p + 2\varepsilon_\ell}\right) r^3 R \frac{V^2}{a^4}\overline{k} \tag{43}$$

The lines of force, or isodynamics, form a family of hyperbolae, x'y' = const. Evaluating (43) per unit volume (of the sphere), we have

$$f = 3\varepsilon_o\varepsilon_\ell KRV^2 a^{-4} \tag{44}$$

and in CGS units

$$f = \varepsilon_\ell KRV^2 / 1.2 10^5 \pi a^4 \tag{45}$$

Here, $K = (\varepsilon_p - \varepsilon_\ell)/(\varepsilon_p + 2\varepsilon_\ell)$, R is the distance between the centre of the sphere and the origin, a is the distance between the origin and the point of intersection of the hyperbola with the x-axis.

e) Translational forces in hyperbolic magnetic field

From the condition of equipotentiality in the gap between two symmetrical hyperbolic pole pieces with upper neutral pole (see Fig. 5) on the surfaces of poles the potential of the field is

$$V_h = A_h R^2 \sin 2\alpha \tag{46}$$

where A_h is the hyperbolic field constant, R is the radius vector, α is the angle between the y-axis and the latter.

In Cartesian coordinates we have

$$yx = \tfrac{1}{2}a^2 \tag{47}$$

where a is the length of the radius at 45°.

In polar coordinates:

$$y = R \cos \alpha \qquad (48)$$
$$x = R \sin \alpha$$

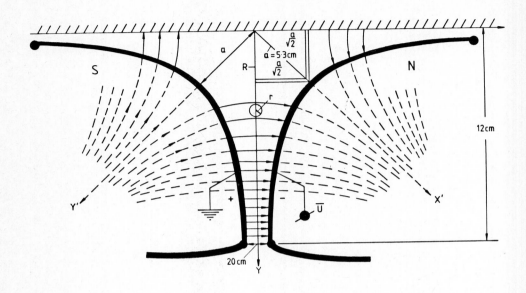

Fig. 5. Spherical particle in hyperbolic polarising fields.

Recalling that for an hyperbole $yx = $ const., we have

$$R^2 . \cos\alpha . \sin\alpha = R^2 \sin2\alpha = \text{const.} \qquad (44)$$

As for the constant A_h, recalling that $\overline{H}$ is given by the gradient of the potential, we have:

$$\overline{H} = \text{grad } V_h = A_h . \text{grad } (R\sin2\alpha) \qquad (50)$$

Denoting by $\overline{k}$ and $\overline{m}$ in unit vectors in the radial and circumferential directions respectively, we have:

$$\overline{H} = A_h (2R\sin2\alpha\overline{k} + R^2\cos2\alpha . 2\frac{\overline{m}}{R}) = A_h (2R\sin2\alpha\overline{k} + 2R\cos\alpha\overline{m}) \qquad (51)$$

For $\alpha = 0$ and $Y = R$, we have:

$$A = H(2R)^{-1} \qquad (52)$$

The central field of the magnetic force generates in the magnetic liquid the additional pressure:

$$P_h = 2(\chi_\ell - \chi_p)A^2R^2 \tag{53}$$

and the magnetic force itself is:

$$\text{grad } P_h = 4(\chi_\ell - \chi_p)A^2R^2\overline{k} \tag{54}$$

For determination of the magnetic force acting on a sphere with radius r on the y-axis at distance R_o from the origin, we use the hydrostatic equation:

$$F_y = -\oint P_h(\overline{e}_y.\overline{n})\,ds \tag{55}$$

where $\overline{n}$ is the vector of the normal to the spherical surface, $\overline{e}_y$ is a unit vector in the y-direction, and ds is a surface element of the sphere. Introducing spherical coordinates r', α', ψ', and recalling that $(\overline{e}_y\overline{n}) = \cos\alpha$, we substitute (53) in (55) and obtain

$$F_y = -\oint P_h\cos\alpha\,dS = -2(\chi_\ell - \chi_p)A^2h\oint R^2\cos\alpha \tag{56}$$

In spherical coordinates, the surface element is given by:

$$dS = r\,\sin\alpha d\alpha d\psi \tag{57}$$

Substituting (57) in (56) we have

$$F_y = -2(\chi_\ell - \chi_p)A_h^2\int_0^{2\pi}d\psi\int_0^{\pi}\left[R_o^2 + r^2(1 - \sin^2\alpha\cos^2\psi) + \right.$$
$$\left. + 2r\,R_o\cos\alpha\right].\cos\alpha r^2\sin\alpha d\alpha \tag{58}$$

Effecting the appropriate integrations, we finally obtain:

$$F_y = -(\tfrac{4}{3}\pi r^3)4(\chi_\ell - \chi_p)A_h^2R_o \tag{59}$$

It should be noted that in this case the force varies linearly with distance from the centre of symmetry.

The volume force is given by:

$$f_y = -4(\chi_\ell - \chi_p)A_h^2R_o \tag{60}$$

We now consider the change of the spacing of the particles with A.   (The latter is varied by regulating the current in the coils of the electromagnet.) The equilibrium coordinates for a pair of particles of different densities, assuming that χ_p' and χ_p'' are small relative to χ_ℓ –

$$R_o' = \frac{(d_p' - d_\ell)\,g}{4\chi_\ell A_h^2}$$

$$R_o'' = \frac{(d_p'' - d_\ell)\,g}{4\chi_\ell A_h^2}$$

(61)

and the vertical spacing is:

$$\nabla R = R_o'' - R_o' = \frac{(d_p'' - d_p')\;g}{4\chi_\ell A_h^2}$$

(62)

i.e it varies inversely as A_h squared.

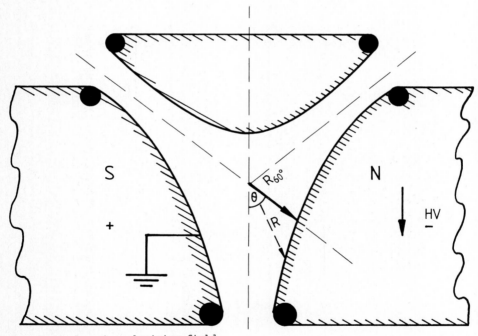

Fig. 6. Isodynamic polarising field.

f) Translational forces in isodynamic polarising fields.

In order to produce the equal electrical or magnetic forces over whole corresponding polarised space the isodynamic configuration should be chosen. Fig. 6 shows these kind of fields.   The profile of electrodes or magnetic pole pieces implys the surfaces of constant potential described by

$$R = R_{60°}(\sin 1.5\theta)^{\frac{2}{3}} \tag{63}$$

The arrangement of magnetic poles of electrodes includes the third neutral piece of same profile. In the case of placing the neutral piece along dashed intermediate axe the profile of neutral piece is simple ortogonal wedge with a spread of 120°.

The squared gradient of electrical field of voltage V in the isodynamic field configuration is

$$\nabla E^2 = \frac{9}{4}\frac{V^2}{R_{60°}^3} \tag{64}$$

So the electrical force acting on unit volume of sphere of radius r is

$$f = \frac{3}{2}\epsilon_0\epsilon_\ell\frac{(\epsilon_p - \epsilon_\ell)}{(\epsilon_p + 2\epsilon_\ell)}\frac{9}{\alpha}\frac{V^2}{R_{60°}^3} \tag{65}$$

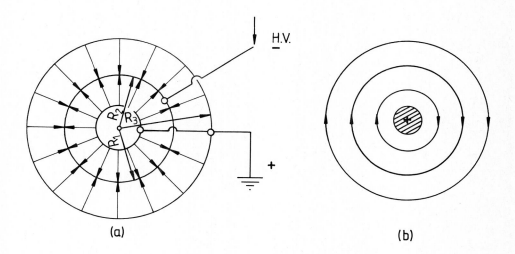

Fig. 7. Polarising fields of cylindrical symmetry.

g) Translational forces in the cylindrical polarising fields.

Translational deflecting forces rising in polarising fields of cylindrical symmetry (see Fig. 7) are convenient for separation purposes. In the case of electrical capacitor the field of cylindrical symmetry is produced by applying

a potential on coaxial cylindrical electrodes. Thus the force lines of the
field intersect the equipotentials at direct angle (Fig. 7(a)).

In the Fig. 7(a) is shown three coaxial cylinders where the potential is
applied to the intermediate one and both internal and external cylinders are
earthed. The square of gradient of electrical field inside this capacitor
for the tori $R_1 - R_2$ and $R_2 - R_3$ respectively are

$$\nabla E_1^2 = \frac{2V^2}{R^3(\ln R_1/R_2)} \tag{66}$$

$$\nabla E_2^2 = \frac{2V^2}{R^3(\ln R_2/R_3)}$$

Here R is the position of particles in the space $R_1 - R_3$.
The expression for the force acting on the sphere of radius r is

$$F = 4\pi\varepsilon_0\varepsilon_\ell \frac{(\varepsilon_p - \varepsilon_\ell)}{(\varepsilon_p + 2\varepsilon_\ell)} \frac{r^3}{R_{600}^3} \cdot \frac{2V^2}{\ln(R_1/R_2)} \tag{68}$$

So the expression for force per unit volume of spherical particles in the tori
$R_1 - R_2$ is

$$f = 6\varepsilon_0\varepsilon_\ell \frac{(\varepsilon_p - \varepsilon_\ell)}{(\varepsilon_p + 2\varepsilon_\ell)} \frac{V^2}{R_{600}^3(\ln R_1/R_2)} \tag{69}$$

In the expression for force rising in the tori R - R the denominator of last
term should be changed for $\ln(R_2/R_3)$.

The above analysed translation forces produce a displacing component of the
force for the formation of separate trajectories of different kind of particles
that is the essence of separation process. The cylindrical magnetic field around
the linear conductor is shown in Fig. 8(b).

4. DIPOLE-DIPOLE INTERACTION AMONG SOLID PARTICLES

As mentioned above, the separation process has a mass nature and usually
takes place in conditions where the distances between particles are of the same
order of magnitude as the particles themselves. Thus the interaction among
particles of a population present in the working space has to be taken into
account. This interaction realise in the rise of attraction/repulsion forces
among particles which is also a result of their polarisation by the external
field. The polarised particles produce their own force fields which are
propogated into the surrounding space and affect neighbouring particles. After
a period of orientation, electrical and magnetic moments of particles align
themselves with the external magnetic field and attraction/repulsion process
takes place.

As an example let us analyse the dipole-dipole interaction in the magnetic
field. For a pair of particles 1 and 2 (Fig. 8) in vacuum, the magnetic force

acting on particle 2 from the induced magnetic field of particle 1 is evidently

$$\overline{F} = (\overline{m}_2 \nabla) \overline{H}_1 \qquad (70)$$

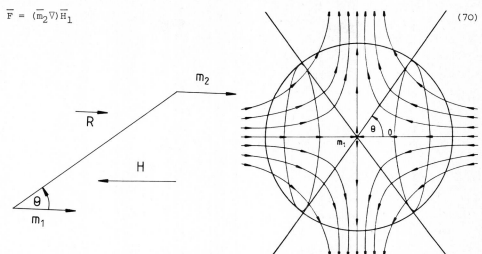

Fig. 8. Interaction between particles in an uniform magnetic field.

It can be easily shown that the magnetic field arising from the dipole of particle 1 is

$$\overline{H}_i = \frac{3(\overline{Rm}_1)\overline{R} - \overline{m}_2 R^2}{R^5} \qquad (71)$$

In the spherical coordinates, the force F (Eqn. 70) has two components (one along $\overline{R}$ and the other along $\overline{\theta}$), and so can be written as

$$\overline{F}_R = \frac{3\overline{m}_1\overline{m}_2}{R^4}(3\cos^2\theta - 1) \qquad (72)$$

and

$$\overline{F}_\theta = \frac{6\overline{m}_1\overline{m}_2}{R^4}\sin\theta\cos\theta \qquad (73)$$

As the magnetisation of a unit volume of a particle is m/V, we can conclude that the interaction forces resulting in particle clusters strongly depend on the distance between particles, magnetic susceptibility of interacting particles, and on the square of the polarising field.   From Eqns. (72) and (73) it is seen that the interaction forces rapidly (proportionally $1/R^4$) decrease with distance between particles.   The existence of the tangential component ($\overline{F}_\theta$) shows that beside attraction-repulsion movements of particles, there is a force causing particles to revolve around one another.

As our experiments on visualisation of particles' behaviour have shown, the period of relocalisation of particles due to their mutual tangential and radial (repulsive-attractive and rotatory) movements results in achieving the equilibrium stationary positions, in the form of clusters.   In the case of the interaction of particles with similar sizes, their equilibrium settlement is complete when all particles are involved in clusters, and chains are formed between opposite poles, along the force lines of the magnetic fields.   In the presence of a sufficient number of particles, chains of clusters become stuck between poles.

In the case of the use of matrix elements (with sizes of 2 - 3 orders of magnitude more than the particles) the equilibrium settlement ends when all particles have found their stationary positions at the attraction areas of matrix surfaces.   Then the friction forces resulting from the attraction pressure on the particles surpass the tangential forces, driving particles around the matrix surfaces.   The end of such a settlement process appears as a building of a deposit of particles on the matrix elements, or on the curvilinear surfaces of pole pieces.

In the absence of a matrix inside the gap between flat pole pieces, instead of these magnetic or dielectric precipitation phenomena, a development of chains of clustered particles takes place.   Both phenomena have a technological significance.   In the case of separation in an electrical field the electrostatic forces due to charges and an inductive image force have to be considered as well.

So, where the components $F_r < 0$, there is attraction between the solid particles, and where $F_r > 0$ there is a repulsion.

For any position of m_2 relative to m_1 a condition exists where $3\cos^2\theta - 1 > 0$, we have $F_R < 0$ and attraction occurs.   As long as $\cos^2\theta > \frac{1}{3}$ and $\cos\theta > 1/\sqrt{3}$, the attraction takes place inside the symmetrical cones of $306^O > \theta > 54^O$ and $126^O < \theta < 234^O$ (see Fig. 8).   Outside this pair of cones a repulsion process occurs

In the case where one of interacting particles is paramagnetic and one is diamagnetic, the product of $m_1 m_2$ in Eqns. (72) and (73) has to be replaced by $-m_1 m_2$.   Then inside the pair of the above designated symmetrical cones a repulsion takes place, while outside these cones there is an attraction.   The particles' interaction in a liquid medium depends on the difference of magnetic susceptibility of particles and medium.   (In the dielectric case it depends on the corresponding difference in dielectric constants.)

The interaction between identical paramagnetic particles in a vacuum, in diamagnetic liquids, and in paramagnetic liquids, with $\chi_1 < \chi_p$, results in chains being produced with coupling inside the pair of cones described above (see Fig. 9(a)).   In paramagnetic liquids with $\chi_1 > \chi_p$, the coupling takes place outside this pair of cones, according to Fig. 9(c).

The interaction between identical diamagnetic particles produces coupling as shown in Fig. 9(b), for all cases when $\chi_1 > \chi_p$, (in vacuum, in paramagnetic media and in diamagnetic media with $\chi_\ell < \chi_p$).   The interaction scheme, shown in Fig. 9(d), takes place in a diamagnetic liquid with $\chi_\ell > \chi_p$.   For two different paramagnetic particles of susceptibilities which are respectively more and less than that of liquid, and for the analogous case of two diamagnetic particles, the configurations of clusters are shown in Figs. 9(e) and 9(f) respectively.   There are four possible configurations for clusters of para- and diamagnetic particles.   The configuration 9(g) relates to the case where the susceptibility of the paramagnetic particle is more than that of liquid and the susceptibility of the diamagnetic particle is less than that of liquid. The configuration 9(h) corresponds to the case where the susceptibility of the paramagnetic particle is less and the susceptibility of the diamagnetic particle is more than that of liquid, whereas configuration 9(j) corresponds to the case where the susceptibility of paramagnetic and diamagnetic particles is more than that of liquid.

Solid dielectric particles in polarised liquid dielectrics produce less varieties of clusters.   There exists only one kind of electrical polarisation - the generation of the internal electrical field in a direction so as to decrease the external polarising field.   For clustering of identical dielectrical particles in different surrounding liquid media, there are two corresponding configurations 9(a) and 9(b);   for clustering dielectrical particles of different permittivities $\varepsilon_p' > \varepsilon_1 > \varepsilon_p''$ configuration 9(f) occurs.

Figure 10 shows a photograph of clustering semi-conducting galena particles in a liquid organic dielectric, with electrical field produced by coaxial round electrodes, and the complete chains of particles developed in the radial direction between the electrodes.

It may be noted that in contrast to magnetic phenomena, which have a purely inductive nature, the processes of orientation, displacement and clustering of particles in the electrical field become complicated by the presence of free charges.   The accumulation of charge on the surface of solids often gives rise to additional Coulombic and image forces, interfering with the pure polarisation phenomena in the behaviour of particles inside the working space of a separator. The image force, arising on the particles as a result of induction by charged electrodes, often overcomes pure polarisation effects.

5. SEPARATING LIQUID MEDIA

a) Dielectric media

Dielectric separation of particulate solids can be performed in a uniform

146

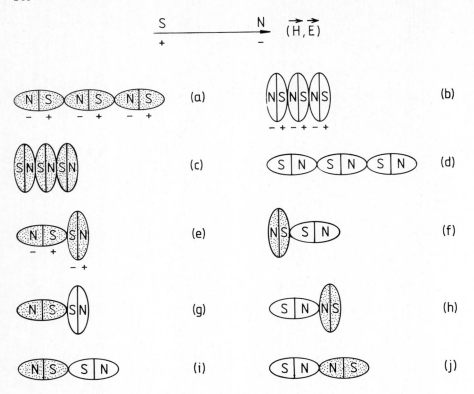

Fig. 9. Clusters of magnetic and dielectric ellipsoids

unimolecular liquids, in mixtures of different unimolecular liquids and in stable
uniform suspensions.   The liquid dielectric medium must posses a combination
of electrical and rheological characteristics which insure the performance of
separation at a technologically acceptable rate.   It is imperative that liquid
should have a minimal toxicity, high environmental persistence, low flammability
and be unreactive in respect of leaching ionic species from solid particles and
the surface of the equipment for separation.

The main performance's requirements are based on the dielectric constant and
electrical strength of media.   So later on we will confine this subject with
discussion of permeability and breakdown threshold of liquid dielectric media.

There is one general rule which may be applied almost to all liquid dielectrics:
the higher its dielectric constant the stronger is the tendency to leach ions
from solids and the higher is its electroconductivity the lower is the breakdown
threshold.

Dielectric separation can be performed in liquid medium of dielectric constant
and of density which are lower than those of solid particulate components to be

Fig. 10. Clustering semiconducting particles in a round electrical field.

separated.   In this case the difference between dielectric constant of solids must be substantial.   Here the non-polar dielectrics of a high electrical strength or their mixtures with polar additives can be used.

In some technological circumstances the separation can be performed in liquids at dielectric constant higher than that of particles.   But of course it is a rare case.

For difficult cases of separation when the components of the mixture have no significant difference in their dielectric constants, liquid dielectrics of intermediate dielectric constant must be used.

Because the wide range of values of dielectric constants of inorganic solids and minerals in particular, for their separation in liquids of intermediate dielectric constants the liquids of practical interest belong to the polar group. However, mixtures of polar and non-polar liquids as well as suspensions of polar solids in electrically strong non-polar liquids can be used.

Although there exist quite simple atomic configurations of molecules with high dipole moments (water, $\varepsilon = 81$, hydrogen dioxide, $\varepsilon = 88$) the majority of substances with high permittivity have a complex microscopic geometry.   The complexity of these molecules makes electrically weak substances.

Permittivity of non-polar and polar substances with relatively simple structure can be calculated on the basis of their refraction indexes and their inter-atomic distances, angles and charge of atoms.   Permittivity of non-polar liquids with small amount of polar additives can be easily calculated on the basis of permittivities of the components.

By introduction or removal of polar groups in the structure of certain
molecules the dielectric constant of the substance can be changed in the
desirable direction.   For instance, the introduction of a second ester group
in a structurally complimentary position into a rigid molecule may increase its
dielectric constant (see Fig. 11).

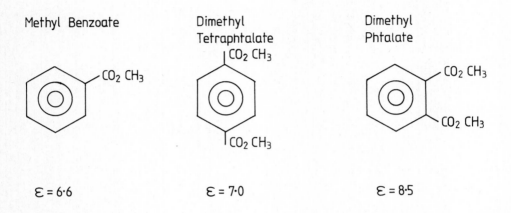

Methyl Benzoate   Dimethyl   Dimethyl
                         Tetraphtalate   Phtalate

$\varepsilon = 6 \cdot 6$ $\varepsilon = 7 \cdot 0$ $\varepsilon = 8 \cdot 5$

Fig. 11. Dielectric constant of phtalates as a function of number and position
of ester group.

In the solutions of different liquids the spatial arrangement of dipole
moments is much more complex than in unimolecular liquids.   So the best way of
determining its dielectric constant is the measurements.

The measurements of dielectric constants of binar mixtures of certain polar
liquids which were made in a coaxial cylindrical capacitor have revealed that
the curve of the function: dielectric constant versus percentage of components,
can have a convexity as well as a concavity.

So there is a possibility of enhancing the dielectric constant of mixtures
beyond the dielectric constant of their components by mixing them in the proportion providing the corresponding spatial arrangement of electrical moments.

b) Two-phase dielectric media

As it was mentioned above, the principle of mixing liquids of high dielectric
strength with polar substances may be expanded for the case of two-phase systems
in which the polar component is in the solid state.   When the solid component
with a high permittivity ($\varepsilon_p \gg \varepsilon_1$) is introduced into a liquid, the resultant
permittivity of the suspension no longer has arithmetic mean of dielectric
constant of the components.   A solid metallic component with zero field inside

increases the electric field in the surrounding liquid phase of the suspension
Proceeding from the Lorenz-Lorentz model of polarisation of an isotropic sphere
in a uniform electric field, the theory of Clausius-Mosotti may be applied to
the permittivity of solid-liquid systems.

Since $\overline{D} = \varepsilon \overline{E}$, we can write for every point of the medium:

$$D = \frac{1}{V}\int \overline{D}(x)\,dv \tag{74}$$

and

$$E = \frac{1}{V}\int \overline{E}(x)\,dv \tag{75}$$

For every point of the medium there exists the equality

$$\overline{D} - \varepsilon_1\overline{E} = \frac{1}{V}\int (\overline{D} - \varepsilon_\ell\overline{E})\,dv = \frac{1}{V}\int v_1(D - \varepsilon_1\overline{E})\,dv + \frac{1}{V}\int v_2(\overline{D} - \varepsilon_1\overline{E}) \tag{76}$$

Here v_1 and v_2 are respectively the volumes of the liquid and solid phases
Inside the liquid $\overline{D} = \varepsilon_1\overline{E}$, hence

$$\frac{1}{V}\int v_1(D - \varepsilon_1 E)\,dv = 0 \tag{77}$$

Then, inside the particles

$$\frac{1}{V}\int v_2(\overline{D} - \varepsilon_1\overline{E})\,dv = (\overline{D} - \varepsilon_1\overline{E})\frac{V_2}{V} = (\overline{D} - \varepsilon_1\overline{E})c \tag{78}$$

For the medium,

$$\overline{D}_m - \varepsilon_1\overline{E}_m = (D_p - \varepsilon_1 E_p)c \tag{79}$$

Since $\varepsilon_m = \overline{D}_m/\overline{E}_m$,

$$\varepsilon_m\overline{E} - \varepsilon_1\overline{E} = \overline{E}(\varepsilon_m - \varepsilon_\ell) = (\varepsilon_p\overline{E}_p - \varepsilon_\ell\overline{E}_p)c = E_p c(\varepsilon_p - \varepsilon_\ell) \tag{80}$$

According to (Lorenz-Lorentz solution for a sphere),

$$\overline{E}_p = \overline{E}\,\frac{\varepsilon_\ell}{2\varepsilon_\ell + \varepsilon_p}\,(\varepsilon_p - \varepsilon_\ell)c \tag{81}$$

From (80) and (81), we have

$$\overline{E}(\varepsilon_m - \varepsilon_\ell) = E\,\frac{3\varepsilon_\ell}{2\varepsilon_\ell + \varepsilon_p}\,(\varepsilon_p - \varepsilon_\ell)c \tag{82}$$

Hence

$$\varepsilon_m = \varepsilon_\ell + c\,\frac{3(\varepsilon_p - \varepsilon_\ell)}{(\varepsilon_p + 2\varepsilon_\ell)} \tag{83}$$

where ε_m is the resultant permittivity, ε_1 is the permittivity of the liquid and ε_p that of the solid component. Equation (83) yields a satisfactory accuracy at low concentrations of solids. (At higher ones, there are additional terms containing the concentration squared.) For $\varepsilon_p \gg \varepsilon_1$, the above equation reduces to

$$\varepsilon_m = \varepsilon_\ell + 3c\varepsilon_\ell = \varepsilon_\ell(1 + 3c) \tag{84}$$

for $\varepsilon_p \ll \varepsilon_\ell$ to

$$\varepsilon_m = \varepsilon_\ell - \frac{3}{2}c\varepsilon_\ell = \varepsilon_\ell(1 - \frac{2}{3}c) \tag{85}$$

and finally, for $\varepsilon_p \simeq \varepsilon_\ell$, to

$$\varepsilon_m = \varepsilon_\ell + c\varepsilon_p - c\varepsilon_\ell = \varepsilon_\ell(1-c) + \varepsilon_p c \tag{86}$$

It is seen that in the first two cases, the resultant permittivity of the suspension (or emulsion) is only slightly affected by the additive, while in the last case the law of additivity is valid.

This calculation agrees with experimental results at a relatively low concentration of the solid component. According to this, the maximum ratio of permittivity is

$$\overline{\varepsilon}_m = \varepsilon_m/\varepsilon_\ell \leqslant 1 + 3c$$

TABLE 1

Relative permittivity of TiO_2 suspension in polystyrene

TiO_2 concentration (%)	Permittivity of suspension		Ratio $\varepsilon_{meas.}/\varepsilon_{calc.}$
	Measured	Calc. (Eqn. 84)	
14	4.35	4.97	0.87
17	5.27	5.28	1.00
19	5.86	5.49	1.06
23	6.80	6.01	1.13
25	7.66	6.12	1.25

Since the suspension concentration for the separation process cannot exceed 0.2 - 0.25, $\bar{\varepsilon}_m$ cannot be higher than 1.6 - 1.75. An example of the permittivity of a two-phase system [1] is given in Table 1, which gives the relative permittivity of a suspension of sintered titanium dioxide powder (ε_p = 125) with particle dimensions of the order of 10 to 10^2 μm in polystyrene (ε_1 = 3.5).

It is seen that in the concentration range of 14 - 23%, the error of Eqn. (84) does not exceed 13%, so that it may be used for a technological evaluation of various media.

c) Ultrafine dielectric suspensions

The theory and experiments described in the Section (b) refer to suspensions in which the dimensions of the solid particles are much larger than those of molecules of carrying liquid.

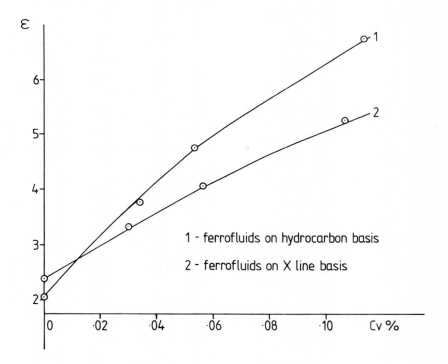

Fig. 12. Dielectric constant of ultrafine suspensions.

For colloidal media with solid semiconducting particles, whose dimension orders are comparable with the dimension orders of molecules of the liquid, the permittivity of the medium at the same concentration is higher than that of a medium of the first kind. Preparation of a suspension with a semiconducting

magnetic solid component is now effected on a commercial scale for solving
different magnetotechnical problems.   For example, ultrafine media of magnetite
particles in water, esters, kerosene, cyclohexane and other liquids are prepared
by protracted (2.0 - 2.5 x 10^3 hours) grinding of ferrite lumps in carrying
liquids, or by chemical co-precipitation of two- and three-valent chlorides of
iron in solutions of ammonium or sodium hydrates.

The stability of such liquids in gravitational, centrifugal, magnetic and
electric fields depends on the ultrafine dimensions, which exceed the dimensions
of the molecules of the liquid by only 1 to 2 orders of magnitude.

Another important condition for stability is that the surface of the solid
particles be covered by 1 or 2 layers of a surfactant like sodium oleate.   For
particles having dimensions of the order of 100 $\overset{O}{A}$, the entropic repulsion exceeds
the molecular attraction as well as the magnetic and dielectric coagulation.

Ultrafine suspensions of magnetite particles in xlene and in isoper were
prepared by protracted grinding in the ball mill.   The density of liquid was in
the range of 790 - 1314 Kg.m^{-3};   the volume concentration of solids was 0.034 -
0.113.   Measurements of the dielectric strength were carried out with electrodes
(two parallel cylinders, d x l = 0.8 x 3.0 cm) at constant voltage.   The
minimum distance between the cylinders was 1.0 cm.   The measurements showed
that the medium did not break down at voltages below 30 kV, whereas individual
discharges took place at 32 - 33 kV with a frequency of about 5 - 10 per minute.
The permittivity of the media was measured in a coaxial cylindrical capacitor
(l = 15 cm, r_e = 3.51 cm, r_i = 3.4 cm) with a capacitance bridge at a frequency
of 1 kHz.   The capacitance of the empty cylinder calculated from the formula
C = 0.2469 $l\left[\log(r_e/r_i)\right]$ in pF was $\simeq$ 400 pF.   Measurement gave 406 pF.

Table 2 gives the results of relative permittivity measurements for a
magnetite-organics suspension at a temperature of 25°C.

Table 2 shows that at relatively low concentration when the equation (84)
gives values of dielectric constant for the coarse suspension slightly (17%)
higher than that of measurement, for the ultrafine suspensions it gives
substantially (2 - 3 times) lower than that of measurement.   This discrepancy
suggests that the pure Lorenz-Lorentz model of polarisation is not applicable
to this kind of system and dipole-dipole interaction should be taken into
account as well as a contribution of the double layer.

6. THE DIELECTRIC STRENGTH OF SEPARATING LIQUIDS
    The upper limit of the applied electric field in dielectric separation depends
on the dielectric strength of the separating liquid, on the one hand, and on the
shape, finish, and material of the electrodes on the other.   The electric
conductivity of the liquid and its dielectric strength depend on some processes
occurring in the electrified medium.   The most important processes are:

Table 2

Dielectric constant of ferrofluids

Ferrofluid	Density Kg.m^3	Concentration of solids %	Dielectric constant, measured	Dielectric constant according to Eqn. (83)	Ratio
Isoper + 5% Oleic acid	790	0.0	2.06	2.06	–
Ferrofluid Saturation 100G	934	0.034	3.75	2.27	1.65
Ferrofluid Saturation 200G	1020	0.053	4.76	2.39	1.99
Ferrofluid Saturation 400G	1270	0.113	6.74	2.76	2.44
Xlene + 5% Enja Y	864	0.0	2.37	2.37	–
Ferrofluid Saturation 100G	984	0.030	3.30	2.58	1.28
Ferrofluid Saturation 200G	1102	0.056	4.01	2.77	1.45
Ferrofluid Saturation 400G	1317	0.106	5.25	3.12	1.68

(1) Electron emission from the metals of the electrodes under the action of an electric field. This surface phenomenon depends on the shape, the surface finish and the work function of the metal.

(2) The phenomena occurring in the liquid and which depend on the electrode separation. They include such effects as the dependence of the conductivity on the viscosity, density and molecular structure of the liquid.

There is a secondary ionization process in the liquid and the influence of any previous dielectric breakdown, which increases the dielectric strength of the liquid. The presence of gases (mainly air and oxygen), adsorbed on the cathode surface and dissolved in the liquid, and impurities (dust) influence the dielectric strength. The dielectric strength of liquids is affected by ionizing agents, the duration of voltage application, the rate of increase of the applied voltage, temperature and pressure.

Measurements show large differences between the electric conductivities and dielectric strengths of low-purity and high-purity liquids. The conductivity of the former is between 10^{-10} and 10^{-8} ohm^{-1} cm^{-1}, while that of the latter can be reduced to 10^{-19} to 10^{-18} ohm^{-1} cm^{-1} by purification processes.

The dielectric strengths of purified hydrocarbon liquids are given in Table 3.

Table 3

Dielectric strength of liquid hydrocarbons

LIQUID	n	M	ρ	E_{bd} (MV m^{-1})
Saturated hydrocarbons $C_n M_{2n+2}$				
Pentane, C_5H_{12}	5	72.14	0.627	144
Hexane, C_6H_{14}	6	86.17	0.659	156
Heptane, C_7H_{16}	7	100.19	0.684	166
Octane, C_8H_{18}	8	114.22	0.703	179
Nonane, C_9H_{20}	9	128.25	0.717	184
Decane, $C_{10}H_{22}$	10	142.27	0.730	192
Tetradecane, $C_{14}H_{30}$	14	198.37	0.762	200
Aromatic hydrocarbons				
Benzene, C_6H_6	6	78	0.879	163
Toluene, C_7H_8	7	92	0.866	199
Ethylbenzene, C_8H_{10}	8	106	0.867	226
n-Propylbenzene, C_9H_{12}	9	120	0.862	250
n-Butylbenzene, $C_{10}H_{14}$	10	134	0.867	275

According to [2], the dielectric strength of straight-chain saturated hydrocarbons is

$$E_{bd} = 2.6 \times 10^4 (n - 4)/(n - 0.427) \, Vm^{-1}$$

For aromatic normal hydrocarbons,

$$E_{bd} = 4.3 \times 10^4 (n - 4)/(n - 0.427) \, Vm^{-1}$$

In Table 3 and here, n is the number of carbon atoms, M is the molecular weight, is the density, and E_{bd} is the dielectric strength (breakdown threshold).

There is one very challenging problem of dielectric separation, i.e. the use of water as liquid dielectric of high relative permittivity ($\varepsilon \sim 81$). The research on dielectric strength of water was carried out in 1961 on the basis of deionization of water in an electric field of about 100 kV m^{-1} [3]. By the deionization the conductivity of water was reduced to 4.3×10^{-6} ohm^{-1} m^{-1}. This is very close to the limit attained in 1894 by Kohlrausch and Heydweiller for water, after 36 successive vacuum distillation stages in gold containers [4]. However, simple deionization of water does not enable us to use it as

dielectric medium in electric fields of the order of 10^3 kV m^{-1}.   It is evident
that the application of another technique for increase of the dielectric strength
of water, eliminating the effect of impurities introduced with the minerals to
be separated should be found.

The University of Grenoble in 1960 started a research programme of producing
low-conductivity and high-permittivity liquids as possible media in electrostatic
generators.   One of the most suitable liquids chosen was nitrobenzene.   The
highest values of the resistivity of purified nitrobenzene, obtained by deionization in an electrodialytic cell, were 3×10^{12} and 10^{11} ohm m in fields of
3×10^3 kV m^{-1} and 3.9×10^4 kV m^{-1} respectively [5].

The separation of mixtures including conducting components alters the conductivity of the separating medium drastically and the presence of conducting
particles creates continuous conducting paths if the threshold of percolation,
which according to [6] is about 15% by volume, is exceeded.

7. MAGNETIC SUSCEPTIBILITY OF LIQUIDS

The magnetic liquids include a large category of fluid media, such as true
solutions of para- or diamagnetic compounts in water or in organic solvents,
liquid oxygen (at low temperatures), liquid organic diamagnetics and, finally,
suspensions of paramagnetic, superparamagnetic and ferromagnetic particles in
various fluids.   To the same category belong fluidized powders of dia-, para-,
and ferromagnetic substances, some of which are described in ref. [7].

A common property of the above media is their interaction with a non-uniform
magnetic field, which induces in them supplementary hydrostatic pressure of
magnetic origin.

Figure 13 shows an hydraulic system with a magnetic liquid inside a solenoid.
Case (a) illustrates the type of liquids drawn into the field along its gradient,
namely, the para-, the superparamagnetic and the ferromagnetic types;   case (b)
illustrates the type expelled from the area of concentration of the force lines,
namely the diamagnetics.   The pressure gradient here is given by:

$$\Delta h = 0.5 (\chi_\ell - \chi_a) (H_1^2 - H_2^2) / (\rho_\ell - \rho_a) g \tag{87}$$

where χ_ℓ and χ_a are the respective volume magnetic susceptibility of the liquid
and the ambient gas, ρ_ℓ and ρ_a are their densities, and H_1 and H_2 are the field
intensities at the initial and final liquid levels.   In accordance with
Langevin's theory, the molal magnetic susceptibility of a solution containing a
paramagnetic and a diamagnetic component may be given by the sum:

$$\chi_\ell = c \frac{N(np)^2}{3k(T + \theta)} + (1 - c) \frac{Ne^2}{6m_oc^2} \sum_{i=1}^{s} r^{-2} \tag{88}$$

where c is the concentration (by weight) of the paramagnetic salt; N is Avogadro's
number (6.02×10^{23} mole^{-1}); k, Boltzmann's constant (1.38×10^{-6} erg deg K^{-1});
n, Bohr's magnetic moment (9.27×10^{-21} erg g s^{-1}); p, the number of Bohr
magnetons; T, the absolute temperature (deg K); θ, the Curie temperature
correction (deg K); e, the electron charge (4.80×10^{-10} esu); m_o, the
electron mass (0.91×10^{-27} g); C, the velocity of light (2.99×10^{10} cms^{-1});
$\bar{r}$, the radius of the orbit projected on the normal plane to the field vector (cm).

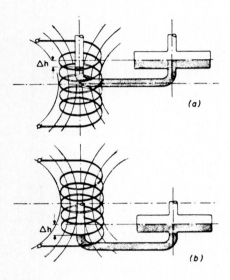

Fig. 13. Behaviour of a magnetic liquid inside a current-carrying solenoid.

As the magnetism of the rare-earth ions considered below is governed by the
spin moments of the deep-lying and screened 4f orbits, their magnetic susceptibility is little affected by the distortive action of the surrounding molecular
fields, or by the interaction of the magnetic-moment carrier ions. In these
circumstances, the correction, θ, may be assumed negligible for these compounds
and the susceptibility relationship is simplified accordingly. The same
applies, with sufficient accuracy, for engineering purposes, for salt solutions
of the iron group.

Accordingly, we have at room temperature:

$$\chi_\ell = 0.4236 \times 10^{-3} cp^2 - (1 - c)\; 0.2382 \sum_{i=1}^{s} \bar{r}^2 \qquad (89)$$

Evaluation of the diamagnetic term is often a problem because of difficulties
in determining the atomic constants of the compound, and in practice, the
measured values given in the literature are often preferable.

In volume terms, Eqn. (89) reduces to the expression:

$$\chi_\ell = 0.4236 \times 10^{-6} p^2 m \rho + (\rho - 10^{-3} mM) \chi_d \tag{90}$$

where χ_ℓ and χ_d are, respectively the overall volume susceptibility of the liquid and that of the diamagnetic component (in emu cm^{-3}), m, the molality and ρ the density (g cm^{-3}) of the liquid; M, the molecular weight of the salt (g mol^{-1})

The type of magnetism of the liquid is determined by the ratio, R, of the two terms:

$$R = 0.42 \times 10^{-6} p^2 m \rho / (\rho - 10^{-3} mM) \chi_d \tag{91}$$

When $R > 1$ the liquid is paramagnetic; $R < 1$, diamagnetic; $R = 1$, non-magnetic.

a) Diamagnetic Liquids

Generation of a substantial magnetic pressure in the diamagnetic liquids is connected with superconducting magnets capable of generating magnetic fields of the order of 10^5 Oe. In such fields, a liquid with $\chi_{dm} \simeq -10^6$ emu undergoes, at a gradient of 10^4 Oe cm^{-1}, a force ($\chi_{dm} \cdot \frac{1}{2}$ grad H^2) of the order of 10^2 dyne cm^{-3}, which often suffices for a separation effect. The practical value diamagnetic liquids increase with increasing of the critical field intensity of superconducting magnets.

Most diamagnetic liquids are organic. The examples in Table 4 were studied at the General Motors Laboratories [8].

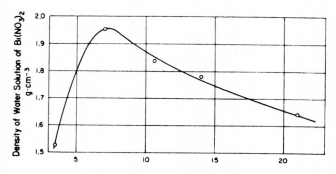

Fig. 14. Saturated aqueous solution of $Bi(NO_3)_3$ - density vs. HNO_3 concentration.

The highest known diamagnetic volume susceptibility is that of metallic bismuth (-9.3×10^{-6} emu), but that of its salts (and especially of their solutions) differs little from that of water. In Fig. 14 the density of the saturated aqueous solution of the nitrate of Bi is plotted against the HNO_3 concentration.

158

At 20°C solubility is maximum at 7% HNO_3, and the corresponding density is 1.95 g cm^{-3}.

The magnetic susceptibility of the aqueous solution of $Bi(NO_3)_2.5H_2O$ with the above density was measured by Gouy's method in the 7 - 16 kOe interval. Fig. 15 shows comparative data for the crystalline salt (-0.92×10^{-6} emu cm^{-3}), the saturated aqueous solution (-0.81×10^{-6} emu), and water.

b) Non-Magnetic Liquids

These, with $\chi = 0$, are the transitional link between the dia- and paramagnetic

Table 4

Diamagnetic susceptibility of organic compounds

COMPOUND	$\chi_{dm} \times 10^6$ emu cm^{-3}
Triiodomethanol	-1.192
Methyl bromide	-1.044
1,2,3-Tribromopropane	-1.023
Ethylene bromide	-0.915
Sucrose solution	-0.877
Water	-0.725

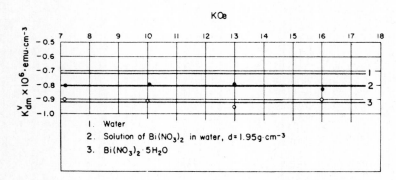

Fig. 15. Saturated aqueous solution of $Bi(NO_3)_3$ - magnetic susceptibility at room temperature.

liquids. They are of interest in experimental and practical engineering problems in which field-medium interaction must be excluded. The simplest way to prepare them is by mixing para- and diamagnetic substances in magnetically equivalent proportions. In Fig. 16 the susceptibility of $MnCl_2.4H_2O$ in aqueous solution is plotted against the molality (1) and density (2) of the solution. The susceptibility is zero at $\rho = 1.0052$ g cm^{-3}, which corresponds to a molality of 0.052.

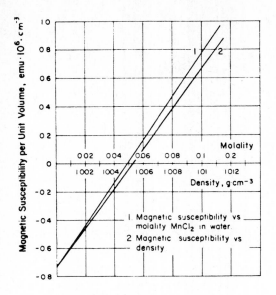

Fig. 16. Dilute solutions of $MnCl_2$ - magnetic susceptibility vs. density/molality.

Table 5

Magnetic moments of paramagnetic metals with p > 5

Serial No.	Atom	Atomic Weight	Relevant Orbit	p
1	Cr	52.0	$3d^5 rs^{-1}$	6.0
2	Mn	55.0	$3d^5$	6.0
3	Fe	55.9	$3d^6$	6.0
4	Ni	58.7	$3d^8$	5.0
5	Co	58.9	$3d^7$	6.0
6	Mo	96.0	$4d^5 5s^{-1}$	6.0
7	Tc*	98.0	$4d^6 5s^1$	7.2
8	Ru	101.7	$4d^7 5s^1$	7.0
9	Rh	102.9	$4d^8 5s^1$	6.0
10	Eu	152.0	$4f^7$	7.0
11	Gd	157.0	$4f^7 5d^1$	8.0
12	Tb	159.2	$4f^8 5d^1$	9.4
13	Dy	162.5	$4f^{10}$	10.4
14	Ho	163.5	$4f^{11}$	10.4
15	Er	167.2	$4f^{12}$	9.4
16	Re	186.3	$5d^5$	5.0
17	Os	190.2	$5d^6$	6.0
18	Ir	193.1	$5d^7$	6.0
19	Am*	243.0	$5f^7$	7.0
20	Cm*	247	$5f^7 6d^1$	5.3
21	Bk*	247	$5f^8 6d^1$	12.0
22	Cf*	249	$5f^{10}$	10.0

* Synthetically prepared

160

c) Paramagnetic Liquids

These are prepared by dissolving the appropriate salts in water or in organic solvents. Their range of susceptibility permits their use for the separation purposes with conventional iron-core electromagnets, generating magnetostatic forces of the order of 10^4 - 10^5 dyne.cm^{-3}.

Of the currently known 100-plus elements, 55 have a magnetic moment [9] and may, in principle, be used in paramagnetic solutions.

Experience has shown, however, that only materials with atoms of p > 5 are suitable for preparing magnetic liquids for mineral separation. (Of these, 22 are listed in Table 5.) An exception is liquid oxygen with p = 2 and χ ranging from 300 to 270 x 10^{-6} emu cm^{-3} in the 55 - 70°K interval [10].

The metals of interest in mineral separation fall under two categories: the iron group (Cr, Mn, Fe, Ni, Co) and the alkaline rare-earth group (Gd, Tb, Dy, Ho, Er), the most representative elements being Mn in the first group, and Dy and Ho in the second. Water-solubility measurements of their salts yielded the following series: sulphide < sulphate < perchlorate < chloride < nitrate.

The solubility of $MnSO_4.2H_2O$, $MnCl_2.4H_2O$ and $Mn(NO_3)_2.4H_2O$ per 100 g water is 85, 151 and 426 g respectively; that of $(Tb, Dy, Ho, Er)_2(SO_4)_3$ at room temperature, 3 - 7 g [11]. Accordingly, chlorides and nitrates were used in the present

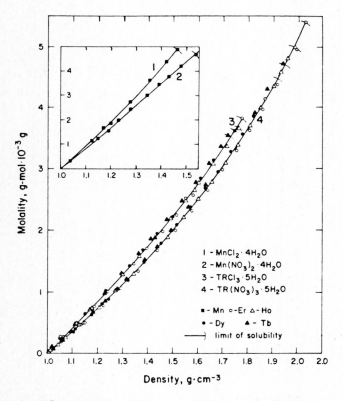

Fig. 17. Magnetic solutions - molality vs. density

study, in which the aspects of interest were three: density versus concentration, magnetic permeability and viscosity vs. density. Solubility was measured pycnometrically, with the salt weighed with an accuracy of up to 10^{-4} g and density determined with an accuracy of up to 10^{-3} g cm^{-3}.

Density is plotted against molality in Fig. 17. For the rare earths, molality was also determined with an allowance made for the bound water (4 or 5 molecules). The limiting solubility of $MnCl_2$ and $Mn(NO_3)_2$ at room temperature corresponds to molalities 4.9 and 4.7 respectively; that of their rare-earth counterparts, to 3.5 - 3.8 and 5.2. The limiting values in the diagram are the maxima.

d) Magnetic Susceptibility

The Faraday method with a quartz cylindrical container of 1.5 cm^3 capacity was used for determining the susceptibility per unit weight in the wedge region of a three-dimensional field (0.8 - 6.0 kOe) generated by conical poles. Gouy's method (quartz tube length 150 mm, inside diameter 5 mm) was used for determining the susceptibility per unit volume in a two-dimensional field (7 - 16 kOe) generated by truncated conical poles. Results were obtained as averages of 5 readings by each method, with the density of the solution under constant control. Scatter did not exceed 5.7%; the average was 2.2%.

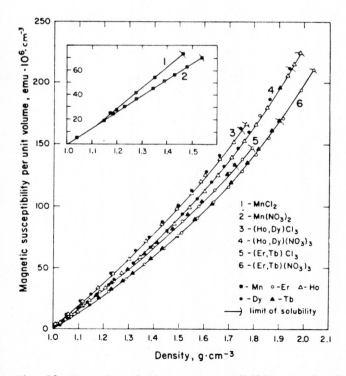

Fig. 18. Magnetic solutions - susceptibility vs. density

The results according to both Gouy's and Faraday's methods are plotted in Fig. 18. The corresponding p values obtained for the ions Mn^{2+}, Er^{3+}, Tb^{3+}, Dy^{3+} and Ho^{3+} in these solutions are 5.9, 9.6, 9.6, 10.3 and 10.3 Bohr magnetons, respectively, in good agreement with the literature data [12]; the susceptibility values obtained from Eqn. (90) on the basis of the above p values are shown in Fig. 18 as solid lines.

A substantial margin for increasing the magnetic susceptibility is provided by working at low temperatures, above the fluidity point. This naturally draws interest to low-solidifying organic solvents. Selwood's paper [13] refers to preliminary studies in this context on alcoholic solutions of $Nd(NO_3)_3$. As was shown experimentally, manganese salts are practically insoluble in organic solvents, while their rare-earth counterparts are freely soluble in alcohol, acetone and propylene carbonate. The latter two crystallize around $200^{\circ}K$, and their solutions of rare-earth salts were studied accordingly. Results for saturated solutions of nitrates of Er, Tb and Dy at room temperature are presented in Table 6. The salts were precalcined at $300^{\circ}C$ with a view to reducing the water of crystallization.

Table 6

Volume magnetic susceptibility of rare earth nitrates in organic solvents at room temperature

	$Er(NO_3)_3$		$Tb(NO_3)_3$		$Dy(NO_3)_3$	
	d ($g\ cm^{-3}$)	$10^6\ \chi_m^V$ ($emu\ cm^{-3}$)	d ($g\ cm^{-3}$)	$10^6\ \chi_m^V$ ($emu\ cm^{-3}$)	d ($g\ cm^{-3}$)	$10^6\ \chi_m^V$ ($emu\ cm^{-3}$)
Acetone	2.013	120.0	1.649	87.8	1.643	108.5
Propylene carbonate	1.755	77.4	1.602	59.3	1.633	74.9

e) Viscosity

Magnetic susceptibility is only one factor in the separation potential of the liquid-solid system, in accordance with the individual susceptibilities and densities of its components. An additional important parameter of a separating solution is its viscosity. The latter, in fact, fully determines the migration velocity of particles at low magnitudes of Re in the working space of the separator. Accordingly, the absolute viscosity of paramagnetic salt solutions was measured on an Ubbelohde capillary viscometer at 25°, assuming Poiseuille flow. Readings were checked also by the relative method against distilled water. Scatter did not exceed 3.2%; the average was 1.4%. The results are plotted against density in Fig. 19. As the viscosity of a solution depends on the kinetic energy of the relevant ions, the dissimilarity of the graphs for chlorides and nitrates is

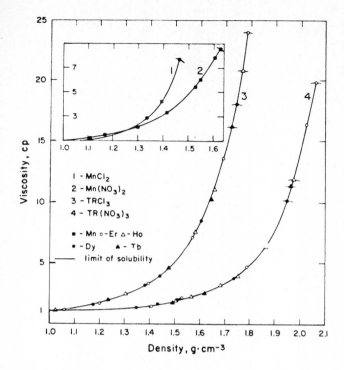

Fig. 19. Magnetic solutions - viscosity vs. density

apparently due to the difference in ionic radius and atomic weight. The viscosity of saturated solutions of Mn salts ranges from 7.5 to 7.7 cp, that of rare-earth salts from 20 to 24 cp. These values must be taken into consideration in separator design in order to ensure sufficient time for the planned migration of the finer particles.

f) Ferromagnetic Liquids

At present, ferromagnetic liquids are known only in binary form, namely as suspensions of fine ferromagnetic particles in diamagnetic liquids. An example of a superparamagnetic suspension used for separation is given in [14]. Because of their high magnetic susceptibility, they are able to generate strong magnetic forces in the relatively weak (<10 kOe) fields of permanent magnets and to be used for the separation process.

The most common version of such a liquid is a suspension of fine particles of magnetite (size 30 - 150 Å), obtained through protracted grinding or joint precipitation of ions of di- and trivalent iron in water or in an organic liquid in the presence of a stabilizer.

The magnetic susceptibility of ester-, hydrocarbon- and water-base suspensions of magnetite, with an amino acid as stabilizer, was evaluated by the Faraday

method in the 0.1 - 8.0 kOe interval, using special spherical vessels with
2.0 cm^3 capacity.   Results (Fig. 20) show that at 0.7 kOe the volume susceptibility has a maximum, namely 70, 78, 98 x 10^{-3} emu cm^{-3}, respectively, for the
above three types of suspensions.   At 7.0 - 7.5 kOe the susceptibility becomes
constant and at equal concentration of magnetite practically equal for the
whole three types, namely 5 x 10^{-3} emu cm^{-3} or three orders higher compared with
manganese salts.   In this context it should be noted that in spite of the
considerable difference between the maximum and the plateau, the working field
must exceed 7.0 kOe for a strong enough expulsive force to be generated.

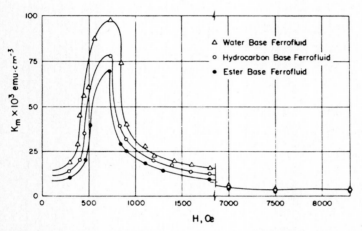

Fig. 20. Magnetic suspensions - magnetic susceptibility vs. field intensity

8. SUPERPOSITION OF POLARISING AND GRAVITATIONAL FIELDS
a) Conditions of separation.

As the separation effect is based on a selective transfer of different
particulate groups into different areas of separators, by electrical and magnetic
forces of a magnitude of order of specific gravity of the particles, the implica-
tion of the gravitational force has to be taken into account.

Where the direction of the magnetic (or electrical) field gradient is strictly
horizontal, the projection of the gravitational force on the plane including the
magnetic or electrical forces is zero.   In this case, the separation of mixtures
can be performed in accordance with the magnetic susceptibility or permittivity
of the mixture components, and independently of particle density.   When the
gradient of the magnetic field is horizontal, the pure magnetic separation of a
mixture of dia- and paramagnetic particles can be performed in vacuum or in air.

The introduction of liquids of intermediate magnetic susceptibility (or
permittivity) into the magnetic or electrical field of the working space of a
separator enables separation to be performed between different diamagnetics, or

different paramagnetics and in the dielectric case, between particles of different permittivity. In these cases, the intersection of the vectors of magnetic (or electrical) and gravitational forces takes place at the direct angle. Thus, the dividing of the mixture only on the two products may be obtained.

Let us direct the gradient of the electrical (or magnetic) field at an angle α ($0^o < \alpha < 90^o$) to the horizontal. Thus, the continuous series of equipotential planes of electrical or magnetic forces will cross the plane of gravitational force at the angle α.

The presence of a liquid medium in the electrical [15] or in the magnetic [16] fields implies the simultaneous rise of Archimedian and electrical (or magnetic) forces, and generates the unique conditions for multifractionation of particulate solids, yielding a number of separation products simultaneously. So, in the gap of the magnet, filled with a liquid magnetic medium (see Fig. 21),

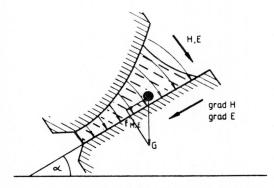

Fig. 21. Solid particles in superimposed polarising and gravitational fields

the net force on a unit volume of particles is

$$f = \frac{1}{2} \text{ grad } H^2 (\chi_\ell - \chi_p) + g(\rho_p - \rho_\ell)\sin\alpha \qquad (92)$$

Here the equilibrium position of the particle will be found at the level where

$$\frac{1}{2} \text{ grad } H^2 (\chi_\ell - \chi_p) = g(\rho_p - \rho_\ell)\sin\alpha \qquad (93)$$

It also implies that

$$\left| \frac{1}{2} \text{ grad } H^2 (\chi_\ell - \chi_p) \right| = \left| g(\rho_p - \rho_\ell)\sin\alpha \right| \qquad (94)$$

The expression corresponding to (94) for the dielectric case is

$$\left| \frac{3}{8\pi} \text{ grad } E^2 \frac{\varepsilon_\ell(\varepsilon_p - \varepsilon_\ell)}{(\varepsilon_p - 2\varepsilon_\ell)} \right| = \left| g(\rho_p - \rho_\ell)\sin\alpha \right| \qquad (95)$$

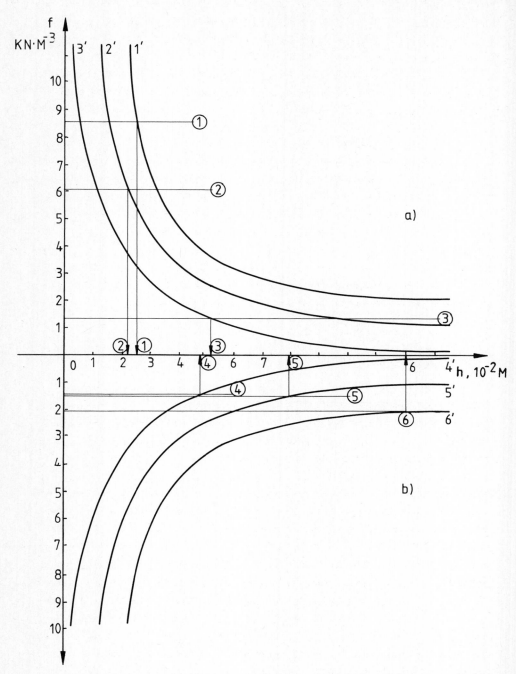

Fig. 22. Positioning particles in superimposed polarising and gravitational fields.

It can be seen from Eqns. (94) and (95) that there are three fixed properties $(\chi_p, \varepsilon_p$ and $\rho_p)$ for given components of a system. The other six parameters $(\chi_1, \varepsilon_1, \rho_1,$ grad H, grad E and $\alpha)$ can be adjusted for effective separation of a particular mixture.

So the choice between magneto-hydrostatic (MHS) stratification in liquid magnetic media, and dielectric medium stratification (DMS) for the treatment of a given mixture, has to be made on the basis of the choice - which process will provide the greatest distances between fractions, stratified in a liquid medium, and ensure the quality of separation.

The simple, graphico-analytical scheme of these processes is shown in Fig. 22. The curves 1' - 6' are produced by calculating the left-hand sides of eqns. (94) and (95) for MHS and DMS respectively. The straight lines 1 - 6 are produced by calculating the right-hand sides of those equations. The lengths and forms of the curves are determined by the choice of profile of the pole pieces or electrodes for MHS and DMS respectively. The position of the curves on the plane f (force per unit volume) - h (height of equilibrium level for a group of particles) can be changed by choosing a separating medium of different χ_1 or ε_1 and by operational changes in H and E. The height of the lines 1 - 6 depends on the density of the separating liquid chosen and can be changed operationally by altering α. The positions of the components on the vertical axis are the prejections of the intersection points of curves 1' - 6' with lines 1 - 6 on axis h.

The quadrant (a) in Fig. 22 corresponds to the case where the magnetic (or dielectric) parameters of the liquid are higher than those of the particles but the density of the liquid is lower. The quadrant (b) refers to the case where the magnetic (or dielectric) parameters of the liquid are higher than those of the particles, but the density of the particles is lower than that of the liquid.

MHS and DMS can be applied to mixtures of components which do not differ in their magnetic or dielectric parameters, but have different densities. The processes are applicable to the case when components have similar densities but differ in magnetic or dielectric properties.

Figure 23 shows examples of stratification by superpositioning of magnetic and electrical fields with a gravitational one. Photograph 23(a) shows the stratification of a three-component mixture of pyrope, zircon and diamond minerals (from bottom to top). Photograph 23(b) shows stratification of a four-component mineral mixture: three mentioned species with an additional fragment of quartz at the top. In this (MHS) case, both parameters of the particles - magnetic susceptibility and density - are responsible for their positions in the magnetic field.

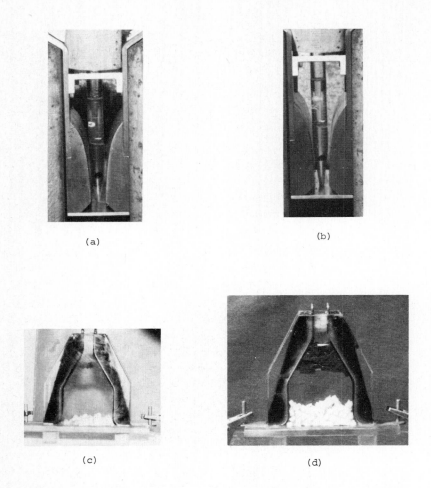

(a)

(b)

(c)

(d)

Fig. 23. Examples of stratification in superimposed fields. (a,b) Magnetohydro-static stratification of a three- and four component mixture respectively. (c,d) Dielectric medium stratification of a two and three-component mixture respectively of plastics.

Photograph 23(c) shows an example of stratification of a two-component mixture of plastics in a liquid dielectric, and photograph 23(d) demonstrates stratifica-tion of a three-component mixture of plastics. In this (DMS) case, the dielectric constant of the particles does not differ and their position in the electrical field is determined by the difference in their density.

In the MHS example, the magnetic susceptibility of the liquid is higher, and the density is lower, than that of the particles. The gradient of the magnetic field is directed from the top to the bottom of the vessel such that magnetic

forces are lifting the particles on the different levels, overcoming the gravitational force.   The separation scheme in quadrant (a) (Fig. 22) corresponds to
this case.

b) Technological aspects

   In the long evolution of technological devices for magnetic and dielectric
separation of particulate solid mixtures, the phenomena described above form a
basis for obtaining desirable metallurgical performances.

   The dominating tendency in the design of commercial and patented separation
equipment was the use of a combination of selective displacement of particles in
a non-uniform polarising field and selective precipitation of particles on
different kinds of magnetic poles, electrodes and matrix elements.   So, in the
variety of designs of magnetic drum separators, the separation is performed as a
precipitation of magnetic particles on the surface of a drum, revolving in a non-
uniform field and providing impenetrable terminal for the solid particles [17].
Pure magnetic precipitations on the surface of magnetized material is used in
the separators with different rigid matrix screens between iron pole pieces [18],
and in periodically working solenoid type of separators, filled with filamentary
matrix as well [19].   The dielectric separator proposed by Hatfield [20] uses
selective precipitation of dielectrics with high dielectric constant on sharp
edges of electrodes itself in the non-uniform electrical field.

   The process of selective coupling of particles in chains is the basis of a
well-known magnetic separator by Davis [21].   In this device the selective
development of magnetic chains takes place against a background of the continuous
demolition of weak diamagnetic quartz clusters, which are washed out by the water
flow.

   The effect of coupling and selective development of vertical chains was reported
in [22] for the dielectric separation of tantalite from quartz and feldspar.

   So the art of designing the separation processes based on the difference in
polarisation of components of particulate solid mixtures is the right choice of
the nature of polarising field, the field configuration, the liquid separating
medium and the mode of displacing mechanism.

REFERENCES

1 A.V. Shubnikov and I.S. Zheludev, Etude des Textures Piézoélectriques, Dunod,
  Paris, 1958.
2 A.H. Sharbaugh, R.W. Crowe and E.B. Cox, Influence of molecular structure upon
  the electric strengths of liquid hydrocarbons, J. Appl. Phys., 27(1956) 806.
3 W. Haller and H. Duecker, Ultra-low-conductivity water, N.B.S. Tech. News
  Bull. (June 1961) 100.
4 W. Kohlrausch and A. Heydweiller, Uber reines Wasser, Z. Phys. Chem., 14(1894)
  317.
5 G. Briere, N.J. Felici and C.F. Filippini, Désionisation du nitrobenzene par
  électrodialyse, C.R. Acad. Sci., 261(1965)5037.

6　I. Webman, J. Jortner and M.H. Cohen, Numerical simulation of continuous percolation conductivity, Phys. Rev. B, 14(1976)4737.

7　M. Filippov, Applied magnetohydrodynamics, Tr. Inst. Fiz. Akad. Nauk Latv. SSR, 1961, p.215.

8　General Motors Res. Lab., Bulletins GMR-317 and GMR-396.

9　H. Kolm, J. Oberteuffer and D. Kelland, High gradient magnetic separation, Sci. Am., Nov.(1975)46.

10　G. Lewis, The magnetism of oxygen and molecules of O_4, J. Am. Chem. Soc. 46(1924)2027.

11　F. Spelding and S. Jaffe, Conductances, solubilities and ionization constants of some rare earth sulfates in aqueous solution at 25ºC, J. Am. Chem. Soc., 76(1954)882-884.

12　Comprehensive Inorganic Chemistry, Vol. 4, Lanthanides, Transition Metal Compounds, Pergamon Press, Oxford, 1st Ed . 1973.

13　P.W. Selwood, Paramagnetism and molecular field of neodymium, J. Am. Chem. Soc., 55(1933)3161-3177.

14　J. Shimoizaka, K. Nakatsuka and T. Fujita, Sink and float separation with magnetic fluid, Private Communication.

15　U.Ts. Andres, Separation of solid particles in liquid dielectrics, Powder Technology, 23(1979)85-97.

16　U.Ts. Andres, Magnetohydrodynamic and magnetio-hydrostatic methods of mineral separation, John Willy, N.Y. (1976)244p.

17　A.M. Gaudin, Principles of Mineral Dressing, McGraw-Hill, New York, 1939.

18　L. Wenz and W. Zabel, Aufbereit. Tech. 13(3)(1973).

19　H. Kolm and A. Freeman, Sci. Am., 212(4)(1965).

20　H.S. Hatfield, Trans. Inst. Min. Metall., 33(1924)335-342.

21　R.S. Dean and C.W. Davis, U.S. Bur. Mines Bull., 425(1941)142-144.

22　G.S. Berger and I.N. Levin, Izu. Akad. Nauk SSSR, (4)(1961)115-117 (in Russian).

HIGH INTENSITY MAGNETIC SEPARATION

R.R.BIRSS and M.R.PARKER
Department of Pure and Applied Physics, University of Salford,
Salford M5 4WT. U.K.

CONTENTS

ABSTRACT

The basic principles of magnetic separation are discussed in detail. This is followed by a review of the design and the mode of operation of a variety of low gradient and of high gradient separation devices, with particular emphasis being given to recently developed systems which use filamentary matrices.

1. INTRODUCTION

The origins of the magnetic separation technique are closely associated with the processing of mineral ores and the history of the subject is fairly well documented. The first recorded patent in this field, applied to the magnetic separation of iron ores, was filed as early as 1792 (ref.1). Most patents and published accounts of magnetic separation relating to the nineteenth century are, in fact, almost totally concerned with the concentration of iron ores. These are dealt with comprehensively in several early twentieth century texts (refs.2,3,4) and the present state of the technology as well as some historical information can be assessed from a variety of books (refs.5-8) and review articles (refs.9-12).

Nowadays, the term magnetic separation embraces a wide range of both design concepts and associated applications. On the one hand, it may refer to relatively simple conventional arrangements such as a lifting magnet (Fig.1) for the removal of ferrous from non-ferrous scrap metal; by contrast, it may also refer to complex laboratory assemblies for the capture of weakly magnetic colloidal particles from cryogenic liquid streams (ref.13) or for the high-field extraction of non-magnetic particles, such as glass (ref.14), from weakly magnetic liquids (Fig.2). It should be made clear, at this point, that a wide range of separation devices are based on the use

of eddy-current forces which are of a magnetic nature. Such devices, which rely on Faraday's law of electromagnetic induction and which are used principally for the separation of metals from non-metals, have been described fully in Volume 1 of this Series (ref.15) and are not dealt with here. Generally speaking, recent developments in high intensity magnetic separation have greatly extended the range of potential application of the technology in areas of vital environmental importance (such as the processing of low-grade mineral ores (ref.8).) In general terms, the fields of application of magnetic separation fall into two main categories. The first of these may be described as applications in which (undesirable) magnetic components are removed from an input stream because the non-magnetic components constitute the valuable product. Alternatively, the technology may be used to concentrate the (valuable) magnetic components in an input stream, the magnetic material being, in this instance, the product of value. Table 1 summarises the commercially established areas of application of magnetic separation; these are numerous, and involve problems in chemical engineering, in the food, drink and tobacco industries and in the manufacturing and metals industries (together with the minerals processing industries, as already mentioned). Table 2 lists a selection of developing applications of high gradient magnetic separation (ref.11) (H.G.M.S.), which is a technique for the removal of

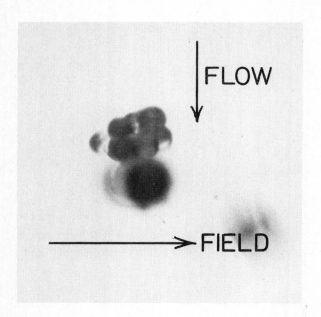

Fig.2. Spherical diamagnetic glass particles being captured on the front face of a magnetized steel wire immersed in a dilute aqueous solution of $MnCl_2$ (ref.14).

of microscopic and weakly magnetic particles from process streams on to capture sites at induced magnetic poles in high fields.

The diverse nature of the information contained in tables 1 and 2 suggests a discussion of separation systems within the framework of some general classification scheme. A variety of schemes are possible (refs. 12,16) such as that based upon device classification in engineering terms (such as belt, drum, isodynamic etc) or upon area of application (such as tramp metal removal, magnetic ore concentration, etc). In what follows, however, the various devices described are classified simply as either low gradient or as high gradient magnetic separation devices.

2. BASIC PRINCIPLES OF MAGNETIC SEPARATION

2.1 Basic parameters

It has been shown (ref.9) that magnetic separation may be regarded as a process of sorting of individual particles of common characteristics based on the competing influences of, first, tractive magnetic forces, secondly, environmental forces (such as gravitation, hydrodynamic drag, friction and inertia) and, thirdly, short-range interparticle forces of various types. These forces add vectorially at appropriate points within the separation system to discriminate between particles of differing magnetic properties in the feed material.

TABLE 1
Established uses of magnetic separation

Device Objective	Areas of Application
Removal of tramp metal (iron,steel, etc)for machinery protection and avoidance of wear.	Chemical and allied industries; food,drink & tobacco industries; manufacturing (paper, glass,plastics, rubber & textiles) industries; household refuse treatment metals processing.
Upgrading of non-magnetic ores by the elimination of reduction of ferrous content(including free iron, magnetite, biotite, muscovite, garnet, iron, silicate, ferrosilicon, ilmenite, tourmaline,horn-blende; in the upgrading of apatite, clay,talc,kaolin, fieldspar,coal,fluorspar,nephelin,baryte, graphite,bauxite, cassierite,etc.)	Mining, quarrying.
Enrichment of magnetic ores (including haematite,magnetite, limonite, Cr ores, Zn,Ni,Ta,Nb,W,Mn).	Mineral dressing.

TABLE 2
Developing Applications of HGMS

Lignite and ash removal from solvent-refined coal;sulphur removal from fuel oil; semi-taconite ore and kaolin clay beneficiation;scavenging from sub-grade deposits or tailings, e.g. mine tips of gold,uranium,platinum, chromium,manganese and iron, and various slags and residues.	Mineral beneficiation.
Removal of heavy metal ions(Cu,Ni, Sn,Pb,Cr.Cd,Fe and Hg) from industrial effluent; removal of TiO_2 from waste water using added Fe++ ions; removal of asbestos fibre from water; clarification of activated sludge in sewage treatment by addition of Fe_2O_3; large scale water purification using magnetic seeding by addition of alum and magnetite; removal of radio-active contamination from PWR thermal reactors cooling systems.	Water treatment.
Recovery of weakly magnetic fine precipitates;magnetic supports for immobilized enzymes in food reactors; red blood cell extraction from whole blood; extraction and concentration of radio-active contaminants in reprocessed fission fuel rods.	Chemical processing.

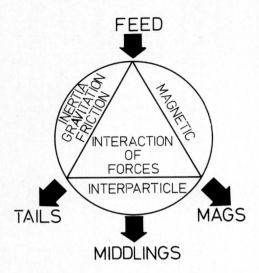

Fig.3. The forces in magnetic separation (after Oberteuffer (ref. 9)

Fig.3 shows a flow diagram, based on the model of Oberteuffer (ref.9) of the working principles of magnetic separators. The feed is divided by the separator into the principal components shown. If the purpose of the device is to isolate a magnetic component, then the non-magnetic exit stream is referred to as the tails. The magnetic component, held in the device, is referred to as the mags and may be removed from the device in its own exit stream. Other product streams containing particles of inter- mediate magnetic properties, referred to as middlings, may also be isolated and removed in certain separation devices. The balance of the various competing forces of Fig.3 determines the comparative levels of magnetic and non-magnetic particles appearing in the mags, middlings and tails, perfect separation being achieved only in very favourable cases.

The efficiency of the separation process may be assessed in terms of either of two parameters. The first of these, the recovery (sometimes referred to as the capture efficiency (ref.17)), is the percentage (or fractional) removal of magnetic material from the feed by the device. The grade is the percentage (or fraction) of the mags that consists of magnetic material. These two parameters provide independent measures of the effectiveness of the separation device ,and their relation to the above-mentioned forces and to one another is discussed in §2.4.

2.2 The magnetic force

Before discussing, in detail, the various forces shown in Fig.3, it is necessary to classify the various substances that can be processed in a separation device. These may be divided into three main types, as illustrated schematically in Fig.4. The first of these are the strongly magnetic (ferromagnetic)materials, in which a

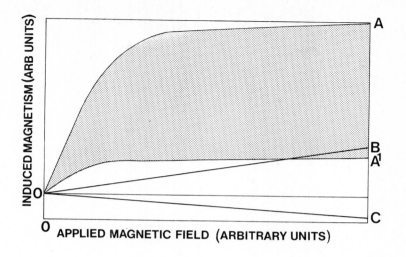

Fig.4. A schematic representation of the response of
strong (A) and weak (A') ferromagnetic, paramagnetic
(B) and diamagnetic (C) materials to applied magnetic
fields. (After Kolm et al (ref.11)).

considerable magnetic moment is induced even by weak magnetic fields.
Such materials quickly reach a saturation magnetization value
which does not alter appreciably thereafter with increasing field.
The value of the saturation field depends on the particular ferromagnetic in question (and partly also on the shape of the specimen
(ref.18)). The second type of material, described as paramagnetic,
is much more weakly susceptible to the influence of an applied
field than are the ferromagnetics. At low field values (in the range
0 to $10^6 Am^{-1}$) its magnetization value may be three or four orders of
magnitude less than those of ferromagnetics. Such materials do not
reach a saturation point even in high magnetic fields and, in fact,
their magnetic moments increase linearly with applied magnetic field.
The third main group of materials, diamagnetics, are essentially non-
magnetic, although they do acquire an extremely small magnetization
of opposite sense to that of the applied magnetic field. This third
type of behaviour may be regarded, superficially, as a negative
(though somewhat weaker) equivalent of paramagnetic behaviour but it
contrasts sharply with the latter in being essentially temperature
independent (ref.18). Diamagnetics, despite their low values of
magnetization even in high fields, may be separated or even captured
in a variety of devices (refs. 5,12) in which their behaviour tends

TABLE 3
Magnetic characterization of the elements (after Kolm et al (ref.11))

1	2	3	4	5	6	7	8	9	10	11	12	13	14	15	16	17	18
H 1																	He 2
Li α 3	Be α 4											B 5	C 6	N β 7	O α 8	F 9	Ne 10
Na α 11	Mg α 12											Al α 13	Si α 14	P 15	S 16	Cl 17	A 18
K β 19	Ca α 20	Sc p 21	Ti p 22	V p 23	**Cr** p 24	Mn p 25	Fe f 26	Co f 27	Ni f 28	Cu β 29	Zn 30	Ga α 31	Ge 32	As 33	Se 34	Br 35	Kr 36
Rb β 37	Sr α 38	Y p 39	Zr α 40	Nb α 41	Mo p 42	Tc p 43	Ru p 44	Rh p 45	Pd p 46	Ag β 47	Cd 48	In p 49	Sn α 50	Sb 51	Te 52	I 53	Xe 54
Cs β 55	Ba α 56	La α 57	Hf α 72	Ta p 73	W p 74	Re p 75	Os p 76	Ir p 77	Pt p 78	Au β 79	Hg 80	Tl β 81	Pb 82	Bi 83	Po 84	At 85	Rn 86
Fr 87	Ra 88	Ac 89	Kh 104														

LANTHANUM SERIES

Ce p 58	Pr p 59	**Nd** p 60	Pm p 61	Sm p 62	Eu p 63	Gd p 64	Tb p 65	Py p 66	Ho p 67	Er p 68	Tm p 69	Yb p 70	Lu α 71

ACTINIUM SERIES

Th α 90	Pa p 91	U p 92	Np p 93	Pu p 94	Am p 95	Cm p 96	Bk p 97	Cf p 98	E p 99	Fm p 100	Mv p 101	No p 102	Lw α 103

to 'mirror' that of paramagnetics.

Table 3 shows a simple form of the periodic table of the elements with the individual elements labelled, in a fashion similar to that of Kolm et al. (ref.11), in respect of their potential for magnetic separation. Here, it is seen that only three elements are ferromagnetic (f) : Fe,Co and Ni. Compounds including one or more of these three elements may be strongly or weakly ferromagnetic or, in some cases, only paramagnetic. Fifty-five elements are seen to be paramagnetic of which thirty-two (p) form compounds which are paramagnetic. A further sixteen (α) are paramagnetic as pure elements but form compounds which are diamagnetic. Another seven elements, (β) become paramagnetic when one, or more, are present in a compound, but two of these, N and Cu, are diamagnetic elements. Of the forty-six remaining elements, thirty are diamagnetic. It is therefore clear that naturally-occurring paramagnetic compounds are far more numerous than those which are ferromagnetic.

The above characterization of materials is relevant to the second major consideration in magnetic separation, namely, how are magnetic forces exerted on substances subjected to applied magnetic fields. The magnetic force producing capture in all magnetic separation systems is the tractive magnetic dipolar force. In general, this tractive force on a particle in vacuo is given by $\underline{F}_m = \nabla \int_V (\underline{M}.\underline{H})\, dv$ (1) (ref.18) where $\underline{M}$ is the magnetization of the particle in the magnetic field $\underline{H}$. The integration is over the entire particle volume. This force is the gradient of the magnetic potential energy of the particle. Equation (1) may be adapted to the case in which the particle is small enough for it to be represented magnetically by a point dipole of moment $\underline{m} = \underline{M}V_p$ where V_p is the volume of the particle. The force on such a point dipole is then

$$\underline{F}_m = \mu_o (\underline{m}.\nabla)\underline{H} \tag{2}$$

If the particle is spherical and non-ferromagnetic with a permeability $\mu_p = (1+\chi_p)\mu_o$ then the magnetization induced in it by a uniform field $\underline{H}$ is

$$\underline{M} = \frac{\chi_p \underline{H}}{1+\chi_p/3} \tag{3}$$

where χ_p is the volume susceptibility of the particle. Combining (3) with (2) it can be seen that the force on a small non-ferromagnetic particle in a uniform magnetic field is

$$\underline{F}_m = \frac{\mu_o V_p \chi_p}{1 + \chi_p/3} \ (\underline{H} \cdot \nabla) \underline{H}$$

i.e.

$$\underline{F}_m = \tfrac{1}{2} \mu_o V_p \chi_p \nabla (\underline{H}^2) \qquad (4)$$

where χ_p is << 1. Finally, if the particle is embedded in a fluid, of volume susceptibility χ_m then the force is modified by the presence of the fluid and assumes the value

$$\underline{F}_m = \tfrac{1}{2} \mu_o V_p (\chi_p - \chi_m) \ \nabla (\underline{H}^2) \qquad (5)$$

(the modification to the force being brought about by subtracting the force on the volume of fluid displaced by the particle).

It is clear from eqn.(5) that the magnitude of this magnetic tractive force is proportional to the product of the magnetic field and the field gradient. The magnitude of the field in separation devices is, of course, limited by technical considerations such as capital and/or operating costs but, within such constraints, near-optimum magnet designs have been produced (refs.8,11). However, a large variety of designs have evolved over the years to produce efficient high magnetic field gradients. Generally speaking, the greater the value of this tractive magnetic force, the more likely it is that the separation will be successful (in that the action of separation can be seen to depend on the relative strengths of a variety of competing forces acting on the particles). These may include, particularly, in the case of small particle systems, hydrodynamic drag forces and, for larger particles, inertial and gravitational forces as well. Separation performance also depends on a variety of short-range interparticle forces existing between, for example, magnetic and non-magnetic particles within the system. The magnetic field and field gradient which act on the particles in separation systems may be engineered in a variety of different ways and this has given rise to a complete range of field intensities and of geometries. Some devices use permanent magnets to produce the field, while others employ current-carrying coils and steel flux -return circuits to magnetize structures the field gradient of which attract the particles.

It is convenient to consider next the part played by the field and field geometry in maximising the magnetic force. Obviously, an increase in the magnitude of $\underline{H}$ increases the magnetic dipolar force both by increasing $|\underline{M}|$ and by increasing $|\underline{H}|$. However, it is clear that the most important aspect is the value of grad.$\underline{H}$, that is how quickly the magnetic field changes with distance.

Although the following argument is a general one, it is of especial relevance to the specific case of the dipolar force on a spherical non-ferromagnetic particle, of radius R, due to the presence of a ferromagnetic cylinder, of radius a , placed transversely in a high uniform magnetic field, $\underline{H}_o$, as shown in Fig.5.   For this arrangement

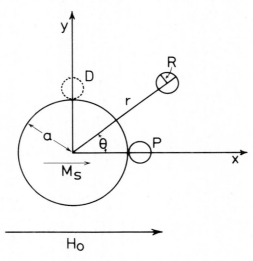

Fig.5. Paramagnetic (P) and diamagnetic (D) particles held at the surface of a ferromagnetic cylinder in the $\theta=0$ and $\theta=\pi/2$ positions respectively.

the radial component of the magnetic field in the vicinity of the cylinder is given by (Appendix 1)

$$H_r = (H_o + \frac{M_s a^2}{2\mu_o r^2})\cos\theta$$

for values of $\mu_o H_o > M_s$.   The radial component of the gradient is therefore

$$\frac{\partial H_r}{\partial r} = - M_s a^2 \cos\theta / \mu_o r^3$$

If $(\chi_p - \chi_m)$ is positive(a net paramagnetic), the particle will experience the greatest force of attraction when in contact with the cylinder in the $\theta = 0$ position (Fig.5).Within the approximation that all of the particle is at the same distance from the cylinder as its

182

centre (that is, assuming that the field gradient does not change appreciably over the particle radius) the magnetic force on the particle is seen to be

$$F_M^P = -4\pi(\chi_p-\chi_m)M_s a^2 (R^3/3r^3)(M_s a^2/2\mu_o r^2+H_o)$$

$$= 4\pi(\chi_p-\chi_m)(M_s H_o/3) \{Kx^4/(1+x)^5+x^2/(1+x)^3\} R^2 \qquad (8)$$

where $K = M_s/2\mu_o H_o$, (9) and where $x = a/R$ is the (dimensionless) ratio of cylinder to particle diameter.

By contrast (ref.14), if $(\chi_p-\chi_m)$ is negative (a net diamagnetic) the maximum attractive force between cylinder and particle exists when the latter is in contact with the cylinder at the $\theta = \pi/2$ position (Fig.5). Here the force is given by

$$F_{Mr}^D = 4\pi(\chi_p-\chi_m)(M_s H_o/3) \{Kx^4/(1+x)^5- x^2/(1+x)^3\}R^2 \qquad (10)$$

Within the constraint that the cylinder magnetization and the applied field have fixed values set by technical considerations then, clearly, the forces given by Eq.(8) and (10) can be optimised, for any given particle size R, at particular values of the ratio, x. The values are dependent on the magnetic constant, K: for example, for K=0.8 x is found to have the optimum value 2.69 for positive $(\chi_p-\chi_m)$ (ref.9) and the optimum value 1.31 for negative $(\chi_p-\chi_m)$ (ref.14).   The variation of F_M, for any given value of R, as a function of the ratio x is shown in Fig.6 using

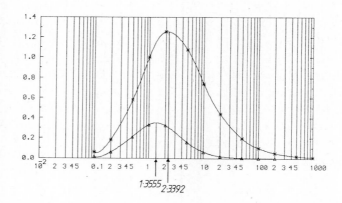

Fig.6. Optimized values of cylinder-particle radii for diamagnetics (lower curve) and paramagnetics. (ref.20)

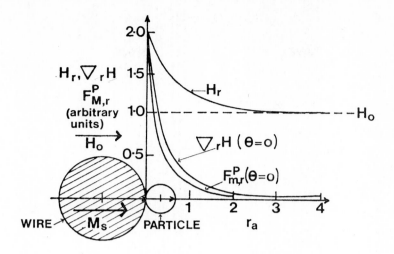

Fig.7. Variation of H_r, $\nabla_r H$, $F_{M,r}^p(\theta=0)$ with wire-particle
distance for the case x=3 (after Oberteuffer ref.9).

a detailed calculation (ref.19) of the optimum value of x (in which
the variation of field and of field gradient over the volume (ref.20)
of the particle is fully considered). This is found to produce
little change in the optimum values for either paramagnetics
(x=2.34)   or diamagnetics (x=1.35).

The general conclusion that may be drawn from this calculation
is that for all magnetic separation devices, optimum performance
is obtained when the spatial variation of the field gradient is
matched to the particle size. This is illustrated schematically,
for the case x = 3, in Fig.7. in which H_r, $\nabla_r H$ and F_{Mr}^p are plotted
along the symmetry direction (0=θ) as a function of distance
between particle and cylinder. This diagram indicates the importance
of the magnetic field having its greatest rate of change across the
diameter of the particle. This may be referred to as gradient
matching (ref.9) and may be regarded as an analogue of impedance
matching in electrical network theory.

2.3 Competing forces
    Equations (8) and (10) express the dependence of the magnetic
force on the particle radius for gradient-matched separation devices.
Clearly, the on-axis radial component dipolar force, in a net
paramagnetic, may be expressed as

$$F_{Mr}^P = -0.854(\chi_p - \chi_m)H_o M_s\ R^2$$

for the case x = 3, so that the force is proportional to the square of the particle radius. For a spherical CuO particle (in water) in an applied field of 10.0T adjacent to a cylinder of saturation magnetization 1.8T, the dipolar force on a 2.0μm diameter particle is approximately 2.91×10^{-10}N. In competition with this force is the hydrodynamic drag force, $\underline{F}_D$, which predominates in the microscopic particle range, For a spherical particle, this force may be generally described, for low Reynold's numbers, by the Stokes expression

$$\underline{F}_D = 6\pi\eta\underline{v}R \tag{12}$$

where $\underline{v}$ is the velocity of the particle relative to the carrier fluid. For a particle of 1.0μm radius moving in water at a speed of $0.01ms^{-1}$, this force is 1.88×10^{-10}N, this value increasing linearly with particle radius, R. The gravitational force, F_g on this CuO particle, of density ρ_p, suspended in water of density ρ_m namely

$$\underline{F}_g = \frac{4}{3}\pi(\rho_p - \rho_m)\ R^3 g \tag{13}$$

is only 2.2×10^{-13}N. Since this force increases in proportion to R^3, it becomes significant only at larger particle sizes. Inertial forces take a variety of forms dependent on the detailed nature of the flow conditions and of the geometrical form of the surface of capture. For example, if the above-mentioned particle is considered as it proceeds on a path of 5.0 m radius (r) around a magnetized cylinder at an angular velocity (ω) of $10^3 rads^{-1}$, the radial (centrifugal) component of the inertial force $\underline{F}_i$ may be described as

$$\underline{F}_i = \frac{4}{3}\pi\rho_p R^3 \omega\underline{r} \tag{14}$$

which gives a value of 1.1×10^{-13}N. Other components of the inertial force may also be of the same order. Clearly, the inertial force, which increases as R^3, may become the predominant competing force over a wide range of particle sizes when the particles are carried through the separation system by a low viscosity(i.e. gaseous)carrier fluid.

Another long-range competing force of less common occurrence, is the eddy current force, $\underline{F}_e$. which operates on highly conducting metallic particles in the presence of a magnetic field. Parker (ref.21) has shown that this may be expressed for spherical particles of conductivity, σ, as

$$\underline{F}_e = (2\pi/15)\sigma \underline{v} (\nabla\underline{B})^2 R^5 \qquad (15)$$

and has the approximate value 4.2×10^{-15}N for a particle of 1μm radius and conductivity $10^7 (\Omega m)^{-1}$ moving with velocity 0.1m s^{-1} in a field gradient of 10^5 Tm^{-1}. Since the gradient must 'match' the particle size, it is clear that this force will increase as R^3, as do gravitational and inertial forces. It has already been demonstrated (ref.15) that this force starts to become important at particle sizes of the order of 1mm diameter.

Another possible long-range competing force is the Lorentz force (ref.18) $\underline{F}_L$ which is independent of particle size and may be expressed as

$$\underline{F}_L = q(\underline{v} \times \underline{B})$$

where q is the electrostatic charge on the particle and $\underline{B}$ is the magnetic induction. This force could possibly be of significance for highly charged particles in gaseous carrier fluids but, as yet, no reports exist of the effects of the action by the force.

The differing rates of variation with particle size of the magnetic force ($\propto R^2$) and of the two major environmental competing forces, of gravitation ($\propto R^3$) and of hydrodynamic drag ($\propto R$), is illustrated in the log-log plot of Fig.8 (ref.20), for a range of particle radius values from 0.1μm to 10^5μm, for the case of a spherical FeO particle moving with a velocity of 0.05ms^{-1} in a water-based separation system with a field of 1.0T. For large particles in a liquid carrier stream, the gravitational force is predominant, although, clearly, at larger sizes. equation (14) shows that inertial forces must be at least comparable in size with gravitational ones. For smaller particles, the viscous drag of the carrier liquid is the predominant competing force. It is obvious from Fig.7 that the magnetic force is dominant in relation to the sum of these competing forces only over a relatively small 'window' of particle sizes (e.g. from about 1μm to 10mm for the FeO particle in the system described above). The width of the window is, of course, dependent upon the net volume susceptibility of the particles and on a number of system parameters (such as magnetic field) and it will obviously be narrower for particles of CuO, for example (ref.9). However, the window is, in any case, probably much narrower than indicated in Fig. 8 because, for smaller particle sizes, inter-particle forces not shown in the diagram become very important, while at the larger particle sizes, inertial forces can

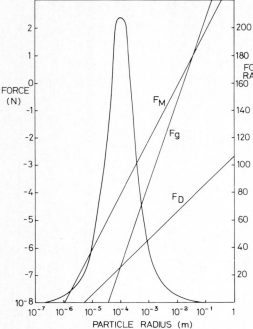

Fig.8. Variation of F_M, F_D and F_g with radius (R) for a spherical FeO particle($(\chi_p - \chi_m) = 7 \times 10^{-3}$ $(\rho_p - \rho_m) = 5 \times 10^3$ kgm^{-3}) as well as the force ratio $F_M/(F_D + F_g)$ (ref.20)

become comparable in influence with gravitation in certain devices. In dry feed systems inertial forces can be extremely important for all particle sizes.

Referring back to Fig.8, it can nevertheless be seen that, for wet separation systems, the condition that the magnetic force is at least equal to the sum of the competing forces is a rough guide to the feasibility of separating the magnetic particles. The ratio of magnetic-to-competing forces, $F_r = F_m/(F_D + F_g)$, for the above-mentioned case of FeO particles is also shown and F_r reaches a maximum value of around 200 for a particle radius of around 100. On the basis of the simple model of Fig.8, F_r can be optimised analytically by differentiating the denominator with respect to R. Here, Equations (12) and (13) show that the optimum particle size occurs when (ref.9) $R = \left[9\eta v/2(\rho_p - \rho_m)g\right]^{\frac{1}{2}}$ (16) irrespective of the magnetic details of the particle. It is clear from equ.(16) that the optimum particle size will be similar for a wide range of magnetic materials (refs.9,20).

2.4 Recovery and grade

The balance between the magnetic force and the competing environmental and inter-particle forces determines the effectiveness of a magnetic separation device, as expressed in terms of the probability of capture of magnetic particles or in terms of the probability that non-magnetic particles are held along with magnetic ones at the magnetic surface.   A detailed understanding of the various short-range inter-particle forces requires reference to specialist texts.   These forces include friction, short-range magnetic dipole-dipole inter-

actions and a variety of charge-dependent forces; these last
include Coulombic, Van der Waals and double-layer forces. In dry
feed systems, larger particles are influenced predominantly by
friction and by moisture-induced forces of adhesion, while very
small particles are almost wholly influenced by electrostatic
interactions. In wet systems containing colloidal particle suspensions,
double-layer forces predominate and the importance of these forces
is strongly dependent on a variety of factors including the pH of the
carrier liquid and the presence or absence of chemical sufactants,
such as Na_2CO_3 or T.S.P.P.

Oberteuffer has shown (ref.9) (with the following analysis)
that the effectiveness of a separation system can be discussed
qualitatively in terms of the forces of Fig.3. From the definitions
of §2.1 the grade of the mags from a separator may be expressed,
in terms of the recovery, R, of the mags and the recovery, R_{NM}, of
the non-magnetic material in the mags, as

$$G = \frac{R}{R + A\,R_{NM}} = (1 + A\,R_{NM}/R)^{-1} \tag{17}$$

where A is the (constant) mass ratio of non-magnetic to magnetic
particles in the feed. It should be noted that R_{NM} and R have
separate functional dependences on the operating parameters of the
system such as flow velocity, separator length, etc. Thus, a single
value of G is obtainable from a range of values of R.

Considering, now, the dependences of R and R_{NM} on the forces of
Fig.3, it has already been suggested above that R is a function of
the force ratio F_r of Fig.8. The simplest possible model is now taken,
in which R is assumed to be directly proportional to this ratio. i.e.

$$R = kF_M/F_C \tag{18}$$

where k is a constant of proportionality, the value of which depends
on system parameters.

The presence of non-magnetic particles in the mags is largely
(though not wholly) due to the action of interparticle forces, F_I,
which reduces the effectiveness of the environmental competing forces,
F_c, in pulling non-magnetics clear of the separator system. R_{NM} is
therefore dependent on the amount of magnetic particles being
captured (i.e. on R) and on the ratio of F_I/F_c. The simplest possible
relationship is one of the form

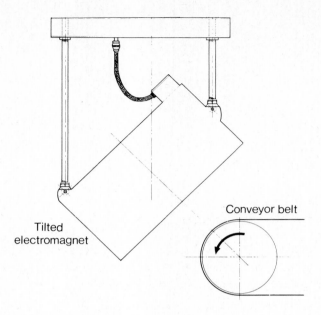

Tilted
electromagnet

Conveyor belt

Fig.9. Overhang magnet (courtesy of
Boxmag Rapid Ltd.)

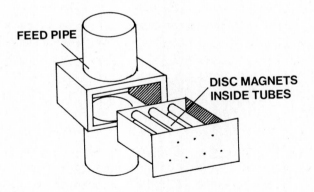

FEED PIPE

**DISC MAGNETS
INSIDE TUBES**

Fig.10. Magnetic grate separator (after
Oberteuffer ref.9)

$$R_{NM} = K^1 R \, F_I/F_c \qquad (19)$$

where K^1 is another constant of proportionality. Combination of equations (17), (18) and (19) leads to

$$G = (1 + Ak^1 F_I/F_c)^{-1} \qquad (20)$$

Equation (20) indicates that, whereas R increases as F_c decreases, G increases as F_c increases. The balance between R and G in a practical separator device is obviously a question of detailed design.

3. LOW GRADIENT MAGNETIC SEPARATION (LGMS)

3.1 Simple LGMS devices

A wide variety of relatively simple devices exists for crude dry separation of ferrous metals, such as the heavy lifting magnet of Fig.1 or the overhang magnet of Fig.9 (designed for the removal of tramp iron from conveyor belts in the processing industries summarized in Table I). A further example is the magnetic grate of Fig.10, comprising, essentially, a series of tubes containing disc-shaped pieces of permanent magnet material of alternating polarity mounted in a pipe through which wet or dry feeds are passed. This device acts as a barrier to the passage of tramp ferrous metal and may be manufactured in a very large range of sizes and types to suit hoppers, floor openings and vertical shutes. A similar arrangement may be used for wet feeds, such as pulp, or foodstuffs, such as soups and juices. Using, say, barium ferrite permanent magnets, it is possible to produce, in this type of device, flux closure fields of 0.05T and field gradients of around 2.6 Tm^{-1}. There are also a number of dry LGMS devices designed primarily for cobbing operations- that is, the rough processing of coarse sands or coarse-lumped ores in order to remove a desired magnetic component. One of these is the dry belt separator, for which a wide variety of commercial designs exist. Photographs and sketches of many of the early commercial designs may be found in the literature (refs.5,6). In most of these devices the feed is supplied from a hopper to a moving conveyor belt that passes through an extended permanent magnet or electromagnet system. Magnetic material, captured by the magnet system, is removed continuously from that system by scrapers or by other means and carried by chutes or secondary moving belts out of the separation device.

One of the most sophisticated devices of this type is the Boxmag OG series of rotating disc separators, illustrated schematically in

(a)

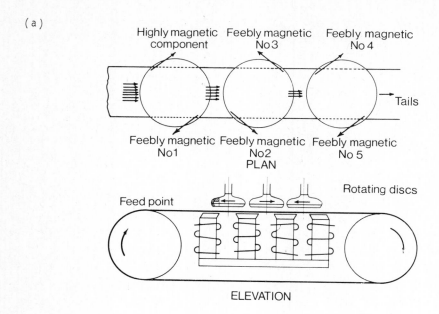

(b)

Fig.11. Dry belt separator (courtesy of Boxmag Rapid Ltd) (a) and photographically in Fig.11(b).

This device is manufactured with a 4(or 3 or 2) pole stationary electromagnet system, and, in intermediate positions above these, 3(or 2 or 1) rotating discs that act as keepers for the magnet system. The revolving discs retain the magnetic particles of the feed and these are then rotated by the discs to points beyond the edges of the discs, where they are removed in zero field by scrapers and discharged on both sides of the belt. The tails are discharged at the second pulley of the conveyor system. The tilt and the elevation of the discs above the belt are adjustable and can be set to optimize arrangements for particular applications. As shown in Fig.11(a), the mags can be subdivided to separate out middlings of intermediate magnetic strength. The capacity of this type of device is somewhat limited, particularly for weakly magnetic ores, but it can be as high as 0.3 tonne hr^{-1} for medium sized particles (-60 mesh) of medium strength ores such as siderite and tourmaline with a belt of 0.38 m width.

3.2 Dry drum separators

The major alternative LGMS system to the dry belt is the dry drum separator. Most modern drum systems are based on the original designs of Ball and Norton (ref.5) the essentials of which are illustrated schematically in Fig.12. Magnetic drum systems are characterized by (usually permanent) magnet poles of alternating polarity arranged in a regular azimuthal distribution close to a limited portion of the inner cylindrical surface. The magnetic field produced by this circular array of magnets extends, typically, throughout a curved region of space from the uppermost point on the outer surface of the drum to a little beyond the diametrically opposite point on the lower surface. This magnet system is surrounded by a continuously rotating non-magnetic cylindrical shell.As Fig.12 indicates, the dry feed enters by means of a hopper at the top of the drum surface. Non-magnetic particles are thrown off by the effects of inertia and gravitation, while the mags are retained at the drum surface by the magnetic tractive force until conveyed beyond the limit of the field. The efficiency of separation is a complex function of the magnetic and other physical properties of the materials of the magnetic field distribution and of the drum geometry and rotation rate. Typical radial magnetic field values for modern drum systems are of the order of 0.05 → 0.1T at the surface. Normally, the magnetic sector of the drum comprises 4 to 6 magnet elements. Laurila (ref.22)

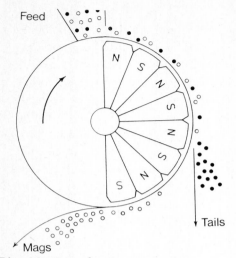

Feed

Tails

Mags

Fig.12. Dry drum separator.

has made a fundamental investigation of the periodic magnetic field of the cylindrical array of magnets with radial poles found in magnetic drum systems. He has found that an optimum pole distribution exists for counteraction of the competing gravitational force at any chosen distance above the drum surface. He has not, however, considered the dynamics of particles moving in close proximity to a drum surface. However, the authors (ref.23) have shown that, for a periodic drum system (Fig.13) , the motion of paramagnetic particles of radius R and volume susceptibility χ_p can be described, ignoring the effects of hydrodynamic drag, by equations of the form

$$m(\ddot{r} - r\dot{\theta}^2) \quad = \quad F_{Mr}, - mg \cos \theta \text{ and} \tag{21}$$

$$m(r\ddot{\theta} + 2\dot{r}\dot{\theta}) \quad = \quad F_{M,\theta} + mg \sin \theta \tag{22}$$

where $\quad F_{M,r} \quad = \quad - (\chi_p R^3 M_s^2/3\mu_o D) (f_r + g_r) \tag{23}$

$$F_{M,\theta} \quad = \quad - (\chi_p R^3 M_s^2/3\mu_o D)(f_\theta \cdot g_\theta) \tag{24}$$

where M_s is the magnetization of the magnetic elements and where $f_{r\theta}, g_{r\theta}$ are periodic radial and azimuthal functions of field. Fig.14 shows compter-generated equipotential contours of the radial component of the field for the case of a drum containing six equal segments of alternating polarity with spacers of zero width (w=0) in Fig.13. It is clear from Eqs. (23) and (24) that a magnetic (or strong paramagnetic) particle will experience a force of varying magnitude and direction as it travels close to the surface. $F_{M,r}$ is always negative (and therefore attractive) and varies only in magnitude, whereas $F_{m,\theta}$ changes sign within each period of the pole pieces   so alternately assisting and then opposing the tangential component of gravity. The periodic nature of $F_{M,r}$ and $F_{M,\theta}$ invariably produces a tumbling motion of the (non-spherical) magnetic particles, which assists their separation from the non-magnetic materia

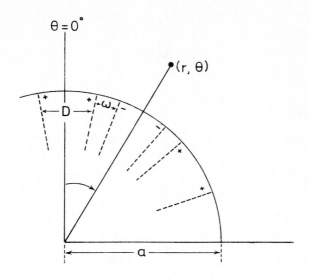

θ = 0°

$(r, θ)$

D

a

Fig.13. Periodic cylindrical magnet
array in a drum separator (ref.23)

A complete analysis of
dry drum behaviour can
only be obtained by
solving individual particle
trajectories of the sort
described by Eqs.(21) and
(22). However, the
following simple analysis
gives a relatively clear
understanding of the
workings of the dry drum
system.

Particles of various
sizes are fed on to the
surface of the non-
magnetic shell around the
θ = 0 position of Fig.13.
The surface of the shell,
usually of polished stainless steel, is almost frictionless, Consider
non-magnetic particles of density ρ_p volume V_p, mass m, in the θ = 0
position. They will require, initially, the (constant) rotating shell
speed   u   and will fall through an angle θ, remaining in contact
with the cylinder so long as their reaction, R, on the surface is
greater than zero. This reaction, being at right angles to the
direction of motion, does no work. Hence the velocity of a non-
magnetic particle at angle θ is given by the energy equation

$$\tfrac{1}{2}mv^2 = \tfrac{1}{2}mu^2 + mga \ (1 - \cos θ)$$

Resolving along the inward normal, at angle θ, gives

$$R = mg \cos θ - mv^2/a$$
$$mg(3 \cos θ - 2) - mu^2/a$$

where the second term on the right hand side is the centrifugal force
described earlier in Eqn.(14). If $u^2 > ga$, R is negative and the
particle is thrown immediately from the drum surface. If, however,
$u^2 < ga$, then the non-magnetic particle leaves the surface of the
rotating shell at an angle.

$$θ = \text{arc cos } \{ (u^2 + 2ga)/3ga \}$$

Magnetic particles, of similar size and density characteristics to the
non-magnetic ones, arriving in contact with the shell surface at the

θ=0 position, will tend to rotate with the angular velocity of the shell. At angle θ, the magnetic particle will remain in contact as long as R ≥ 0. Thus

$$R = F_M(r,\theta) + mg \cos \theta - mu^2/a$$

where $F_M(r,\theta)$ is the (periodic) magnetic traction force attracting the particle to the surface (Eqs.(23) and (24)). Clearly, a satisfactory separation of highly magnetic middlings and tails can be arranged if u is made large (causing the tails to leave the surface at small values of θ) and if F_M is made sufficiently large (such that R is just equal to zero at θ = π say). For this latter condition to hold

$$F_M(r,\theta) = mg + mu^2/a \qquad (25)$$

For a ferromagnetic particle, the force term may be represented approximately in two ways. First, if the average field value of the permanent magnet array of Fig.13 is not very large (say 0.04T) and if the geometry of the particle indicates that a large demagnetising

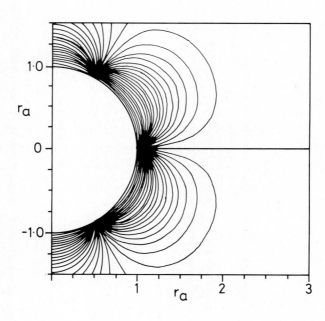

Fig.14. Calculated radial magnetic field
contours of a drum system (ref.23)

field is needed, then the ferromagnetic particle will exhibit
behaviour akin to a very strong paramagnetic one of volume
susceptibility 3.0. F_M is then

$$\frac{3}{2} V_p \mu_o \nabla(H^2)$$

and equation (25) becomes

$$\mu_o \nabla(H^2) = \frac{2}{3} \rho_p (g + u^2/a) \tag{26}$$

Secondly, if the field and particle shape are such that the particle
approaches or exceeds its saturation moment, then F_M can be expressed
as

$$\mu_o V_p (M \cdot \nabla)\underline{H}$$

and equation (26) becomes

$$\mu_o (\underline{M} \cdot \nabla)\underline{H} = \rho_p (g + u^2/a)$$

Consider a dry drum separator of 0.5m diameter, with (average) field
and field gradient at its surface of 4.0×10^{-2}T and 4.0×10^{-2} Tm^{-1},
respectively. The resultant magnetic tractive force on a spherical
steel particle (equation (4)), of 100µm radius, would be 1.25×10^{-7}N
compared with a gravitational force of 2×10^{-7}N and a centrifugal
force of 6×10^{-7}N, for a drum rotation rate of 100 rpm. Obviously,
this rotation rate can be adjusted to enable weakly magnetic
materials to be extracted at convenient intermediate values of θ.

Dry drum separators have been found to be more effective on coarse
feed materials, with particle sizes typically larger than 6mm diameter.
They are limited in performance by a variety of technical factors, such
as the particles being physically too small or magnetically too weak,
or the middlings being insufficiently detached from non -magnetic
'gangue' material in the feed. The feed is, of necessity, almost
totally dry and 'screened' into a selected size range. It must be
dust-free and introduced into the machine at a depth of about one
monolayer. The efficiency of this type of device is reduced by the
tendency of crushed or milled dry particles to adhere to one another
and to surfaces through charge-dependent forces. However, as in the
case of the dry-belt and the induced roll (see later) separators,
this type of device will always be important in hostile environments
where geographical (e.g. desert) or climatic (e.g. arctic) factors
make wet technology inappropriate.

3.3 Wet drum separators

Most present-day wet drum separators are based on an early design by Grondal (ref.25) and employ the 'pick-up' principle in which magnetic particles are lifted from the wet feed while the non -magnetic gangue streams through.

This is currently the most widely used magnetic separation device in the world, primarily in connection with the concentration of vast tonnages of low grade iron ores. It generally consists of a cylindrical drum containing, typically, between 3 and 7 fixed magnet poles arranged in a manner similar to that shown in Fig.13 but restricted to an angular sector in the lower part of the drum roughly defined by $\theta = 135^{\circ}$ and $\theta = 225^{\circ}$ in Fig.13.

Surrounding this magnet system there is usually an austenitic stainless steel shell mounted on corrosion-resistant aluminium, phosphor bronze or stainless steel end flanges, mounted in turn on self-aligning sealed ball bearings. Double seals ensure watertightness and the drum is mounted within a stainless steel tank. Above the tank is a 'header box' - a two-component spillover vessel in mild steel with a stainless steel base. A weir arrangement ensures that a uniform feed to the drum is obtained and a further separate weir system eliminates the effects of surges in the feed system. The wet feed is either pumped to, or gravity-fed to, the header box. A rubber scraper and wash spray pipe are usually installed for the removal of magnetics, although the latter is not suitable for heavy media recovery applications (e.g. coal cleaning) because of the need to maintain high specific gravity. The particle size in the wet slurry feed is typically less than 6mm to avoid practical problems assocated with clogging of the device.

Magnetic ore concentration often takes place in three distinct processing stages. The primary stage is known as 'cobbing' and here the principal objective is high recovery of the magnetics including those attached to gangue particles. Secondary crushing of these ores usually liberates more magnetics and so a second 'roughing' stage is used in which the feed is more highly magnetic than at the cobbing stage. Further grinding of the ore usually releases additional small amounts of gangue and this is discharged as tails in a final 'finishing' stage. Separation performance is crucially dependent upon the design of the separator 1box . The particular design adopted depends on factors such as the manner in which the feed is brought to the screen surface, the removal of gangue and the details of the washing action. However, it is possible to subdivide most commercial drum systems into three main classes, namely concurrent, counter-rotation and counter-current (ref.26). Schematic outlines of the three main system types are shown in Figs.15(a) (b) and (c).

In the first of these (the concurrent flow system) the wet feed enters at a pre-set level and flows in the same direction as the drum rotation. This is, typically, a high capacity system, used primarily as a cobbing stage for coarse ground ores of particle size less than 6mm. The tails are easily removed from this device and, for relatively coarse ores, this type of device has a capacity (processed weight/unit width of drum) of about 30-50 tonne m^{-1}. The counter-rotation type (Fig.15(b)) usually follows secondary crushing of the ore in a ball-mill system in which more magnetics are liberated. This is also a high capacity device in which the cleaning of the magnetic concentrate is of only secondary importance. The capacity of this type of device for feed ground to less than 14 mesh(1.0mm)is around 50-65 tonnes m^{-1}. The counter-current device is used mainly for finishing, where very thorough cleaning of the concentrate is required. The name derives from the fact that the tails are forced to flow (Fig.15(c)) counter to the rotation of the drum when leaving the machine. This design provides a high grade concentrate with minimum losses of fine concentrate. The slurry is fed into the device directly beneath the drum and any mags washed free of the magnetic field during the powerful crushing process go back into the cleaning system and are re-processed. These devices are mostly suited to finely ground material, a typical system capacity at -100 mesh (150μm) being $6 \rightarrow 10$ (dry) tonne m^{-1}.

Commercial versions of wet drums manufactured by Boxmag-Rapid Ltd. are made in 4 different diameters, 0.6 m, 0.75 m, 0.9 m and 1.07 m and in cylinder widths of up to 3.05 m. Drum speeds are commonly set at around 20 rev/min. The two primary uses of wet drum separators throughout the world are those of heavy media recovery and of iron ore concentration.

In commercial dense media recovery systems, the drums are required to handle a wide range of feed concentrations, from dilute feeds from screen filters to highly concentrated feeds from cyclone

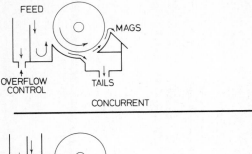

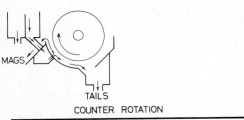

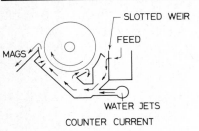

Fig.15. Wet drum separators (after
    Suleski ref.26).

separators. Fig.16 shows one such large commercial installation of wet drums for the extraction of magnetite from phosphate ore.

3.4. Wet belt separators

This type of device is used mainly for high capacity concentration of iron ores or for magnetic recovery in dense-media separation. Modern devices retain most of the characteristics of a 1935 design by Crockett (ref.27). Its construction is shown schematically in Fig.17, from which it can be seen that a rubber belt, approximately in the shape of a catenary, running between two pulleys, dips at its lower extremity slightly beneath the surface of wet feed (supplied at about 30% solids concentration) to a trough above which is an alternate-pole magnet system. The mags are carried forward on the lower surface of the belt under the influence of the magnetic field while the gangue settles at the bottom of a large trough, whilst the mags pass beyond a weir and leave the belt at the extremity of the magnet system. This device is best suited to the separation of magnetic particles greater than 100 mesh (150μm). At smaller particles sizes, it can not compete effectively with wet drum separation. The main advantage of this device, besides its low requirements on electrical running power, is that it provides a relatively dry magnetic concentrate in a single passage. The machine usually separates using a belt width of around 1.0m; the overall height of the machine is around 2.5 m. At this size, the capacity is around 10 tonne hr^{-1}. This is a robust device, needing little maintenance, which is capable of handling wet feeds of widely varying consistency. When used for dense media applications, the residual magnetism in the mags is removed, usually by magnetic de-flocculation in an AC coil,

Fig.16. Installation of two WCF 2427 Concurrent
Finisher Concentrators separating magnetite
from copper phosphate ore (Courtesy of
Boxmag Rapid Ltd.)

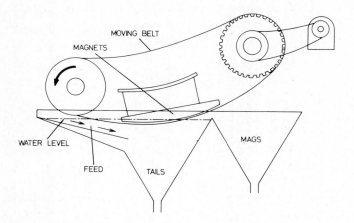

Fig.17. Schematic of wet belt separator
prior to its return to the dense media bath.

4. HIGH GRADIENT MAGNETIC SEPARATION (HGMS)

Dry HGMS techniques have been used industrially for some considerable time for the concentration of the less strongly magnetic iron-
bearing minerals, such as weakly magnetic sands, haematite and other
hydrous oxides of iron. The problem presented by the combination of
the relatively low volume susceptibilities of weakly magnetic minerals
and small particle sizes cannot be solved by drum and belt operations.
The most widely used of the established commercial devices in this
field of operation are the induced roll separator (ref.28) and the
Frantz Ferrofilter (ref.29),(the latter also being useful for wet
separation).

4.1 Induced roll separator

The modern form of this device, loosely based on a design
innovated by C Q Payne as long ago as 1908 (ref.28), is outlined
schematically in Fig.18. Here, the mineral feed, in the form of
a fine dry powder, is released at a controlled rate from a feed
hopper or a vibratory feeder directly on to a magnetized separating
roll (or system of rolls) which carries the powder through a magnetic
field between the roll surface and an adjustable magnet pole piece.

Magnetically susceptible particles in the feed are attracted to,
and held at, the magnetized surface of the roll. The rotation of

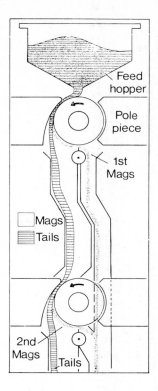

Fig.18. Schematic of induced roll separator (courtesy of Boxmag-Rapid Ltd.)

the roll carries these particles beyond the so-called splitter plate, where the field gradient at the roll surface reaches a minimum and where the particles are discharged with the aid of a brush. Particles of low susceptibility are discharged earlier because the centrifugal force imparted by the rotation of the roll overcomes the magnetic traction force acting on the particles. The separating roll comprises a non-magnetic stainless steel roll shaft around which are fitted annular ferromagnetic discs coupled together with (magnetically) insulating spacers, such as aluminium, in a laminar assembly. These laminations reduce eddy currents to negligible proportions and, at the same time, interrupt the field gradient at the roll surface. The profile of the adjustable left-hand pole piece of Fig.18 can be any one of a variety of designs, depending on the particular application. The roll speed is variable and can be adjusted to a value somewhere in the range 80 to 800 rpm to optimize performance. Commercial laminations are of the order of 1.0 mm and magnet fields of the order of 2.0T are employed, resulting in field gradients of the order of $10^3 Tm^{-1}$. Considerable variations in throughput are found for any given device, depending on the material and on the particle size being processed. Fig.19 shows a Stearns (Appendix 2) 30-inch 3-Field KT model induced roll separator used for the recovery of staurolite from heavy mineral concentrate in Florida. Table 4 shows some typical processing rates quoted by Boxmag Rapid Ltd. (Appendix 2) for a 0.75 m wide roll. Table 5 summarizes the principal technical parameters of the Boxmag Rapid range of industrial induced roll separation systems.

4.2. Frantz Ferrofilter

This is an industrial separation device manufactured by S G Frantz Company Inc. (Appendix 2) consisting of a regular array of ferromagnetic stainless steel screens or grids arranged throughout the field volume of a solenoid magnet (Fig.20). The screens are ribbon-shaped and are

Fig.19.   30 inch 3-Field KT model induced
roll separator (courtesy of Stearns
Magnetics Inc.)

arranged with their edges parallel to the (axial) magnetic field in
order to facilitate the saturation magnetization condition. The
Frantz Ferrofilter is a flexible device which can operate successfully
on wet or dry feeds and which has the capability of separating out
colloidal particles of strongly magnetic materials. The solenoid
produces only a modest field (0.2T) but this is sufficient to saturate
the flat ribbons of ferromagnetic material and to induce poles along
the ribbon edges. The force density on colloidal iron particles at
these ribbon edges is around $10^9 Nm^{-3}$ (ref.30) implying field gradient
values of the order of $600Tm^{-1}$. An evaluation of the various

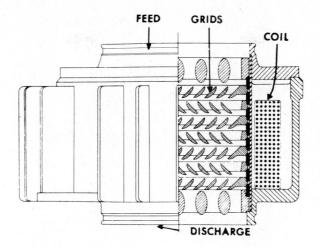

FEED GRIDS COIL DISCHARGE

Fig.20. Frantz Ferrofilter (courtesy of
S.G. Frantz Company Inc.)

competing forces on colloidal ferromagnetic particles indicates that
it is possible to pass the feed through the device at speeds in the
range 10-50mms^{-1}. The ferrofilter is an intermittent device in
which the capture sites on the screens eventually become saturated
with magnetics, whereupon the flow is interrupted and the device
washed clean in zero magnetic field. The essential characteristics
of the field pattern of the regular array of ribbon-like screens
used in this type of device can be found from Eqn.(44) in §5.1
below.

The most widely used of the well established forms of wet HGMS is
the Jones Separator which is marketed on a worldwide basis by
British and West German manufacturers (Appendix 2).

4.3 The Jones separator

In this device, first patented by Jones in 1955 (ref.31), a
thoroughly mixed slurry, containing no particles larger than 1.0mm,
falls vertically through feed pipes between the poles of a strong
electromagnet containing a series of sharply grooved plates in the
field space (Fig.21). In this type of system, with grooves of
around 1 mm depth, and a background field of 2.0T, field gradients
of the order of 2.0 x 10^{3}Tm^{-1} are obtainable. Such gradients enable
the effective separation of slurries of up to 25% solids content for

TABLE 4

Processing Rates for a 0.75m width roll (Boxmag Rapid)

Application	Particle Size Range	Processing Rate thr^{-1}
Purification of quartz	-10mesh(1.67mm)	6
Purification of abrasives	-14mesh(1.2mm)	3
Extraction of ilmenite from heavy mineral sand	-22mesh(0.7mm)	3
From schulite purification of rutile and zircon	-22mesh(0.7mm)	3
Separation of wolframite from schulite	-150mesh(0.105mm)	1.5
Separation of siderite from cassiterite	-170mesh(0.088mm)	1.5

TABLE 5

Technical parameters of large-scale induced roll separators (Boxmag Rapid)

Model	Type of Device	Net Weight (kg)	Motor (kw)	Coil power (kw)	Roll width (mm)
IRB 1.250	1 roll	1465	0.5	1.1	250
IRB 1.500	1 feed point	2490	0.75	1.5	500
IRB 1.750	Single stage unit	3290	1.1	1.9	750
IRB 2.250	2 roll.1 feed point double stage	2040	1.1	1.75	250
		3450	1.5	1.9	500
IRB 2.750	unit	4551	2.2	2.4	750
SHR 500.1	2 roll 1 feed	3750	1.5	1.9	500
SHR 750.2	paint single stage unit	5350	2.2	2.8	750
SHR 500.2	4 roll feed points	7000	2x1.5	3.5	500
SHR 750.2	single stage unit	9950	2x2.2	5.0	750
DHR 500.1	4 roll,1 feed	5000	3.0	4.4	500
DHR 750.1	point, double stage unit	7300	3.0	5.7	750
DHR 500.2	8 roll, 2 feed	9550	2x2.2	6.2	500
DHR 750.2	points, double stage unit	13550	2x3.0	9.2	750

a wide variety of minerals of intermediate magnetic strength. In large commercial Jones or marketed Jones-type separators, such as those marketed by Boxmag Rapid Ltd, the slurry can be fed continuously to the device due to rotation of the grooved plate boxes or array of steel bars on rotor systems (Fig.22). Within the magnetic field, the grooved plates concentrate the magnetic flux at the tips of the ridges (inset Fig.21). The gap between adjacent

plates is adjustable. Within the field, feebly magnetic particles
adhere to the plates, while the non-magnetic pulp passes straight
through the plate boxes and is collected beneath. Before leaving the
magnetic field, entrained non-magnetic particles are washed out by
low-pressure wash-water. Washings of this type are usually
classified as middlings and are collected separately. When the plate
boxes reach the 'demagnetized zone' (ie a zone roughly midway between
the two magnetic poles, the magnetic particles are washed out with
high-pressure scour water sprays operating at about 50tm^{-2} (~ 70 psi).

To promote a high recovery in the Jones device, it is necessary to
mill or screen the wet feed to ensure that the mags are liberated.
Where the feed has a significant proportion of very small particles
it is sometimes necessary to add a dispersant to the feed. Sometimes,
stringent product quality requirements necessitate a double pass of
the wet feed through the grooved plate system.

Table 6 shows operating parameters for a wide range of commercially
available rotary versions of this device produced by KHD Industriean-
lagen AG (Appendix 2). The machine capacities (in thr^{-1}) vary
considerably with the nature and identity of the wet feed slurry.
Those shown in the Table are for finely divided Brazilian haematite
ore.

Jones (ref.32) has shown that the field produced from each ridge
of salient groove plate can be considered as that of a line dipole
at the centre of the grooved ridge radius. This radius is h/4 if **h**
is the groove pitch (Fig.23). The field system of a regular array
of line dipoles has been treated by a variety of authors (refs.33-37)
and is dealt with fully in §5.1(see equations (44) and (45)). However,
a nearest-neighbour calculation, as undertaken by Jones, (ref.37),
is simple and effective in allowing an understanding of the relative
magnitudes of the magnetic and competing forces. Consider the pole
plate geometry of Fig.23. From eqn.(8) of §2.1 it can be seen that
the tractive magnetic force upon an 'on-axis' particle at the point
O of Fig.23. is expressible as

$$F_m = \mu_o \chi V_p M_s a^2 H_o \left(\frac{1}{r_x^3} - \frac{1}{r_y^3} \right) \tag{27}$$

where a(=h/4) is the effective radius and r_x, r_y are the respective
distances of O from X and Y, For a measured mid-point (S) magnetic
field of 1.6T and for r_1 = 1.0mm and r_2 = 1.8mm, Jones calculated the

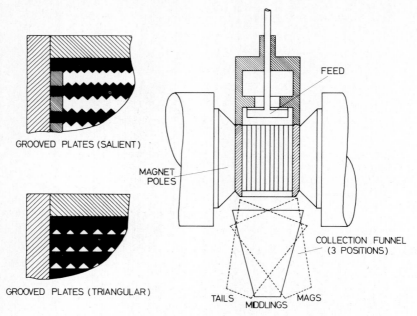

GROOVED PLATES (SALIENT)

MAGNET
POLES

FEED

COLLECTION FUNNEL
(3 POSITIONS)

GROOVED PLATES (TRIANGULAR)

TAILS   MAGS
MIDDLINGS

Fig.21. Schematic of the Jones Separator

net-pull on the particle towards X as $9.2(\chi_p-\chi_m)V_p \times 10^3$ N. For a 52 mesh (0.295mm diameter) spherical particle of siderite (χ_p = 4.3 x 10^{-4}, ρ_p = 3.8 x 10^3 kg m^{-3}), this force is around 110 times the weight of the particle. Within this simplified approach it is possible to calculate, in a similar fashion, the net sideways pull on an particle at an arbitrary off-axis position, R, as shown in Fig.23. Here, from eqns.(4),(6) and (7) of §2.1 it can be seen that this force is expressible as

$$F_m = \mu_o(\chi_p-\chi_m)V_p \left(\frac{\sin \theta}{r_1^3} - \frac{\sin \phi}{r_2^3} \right)$$

where θ, ϕ, r_1 and r_2 are defined in Fig.23. For the same siderite particle and for ϕ = $\pi/6$, r_1 = 1.61mm and r_2 = 2.8mm, the sideways thrust on the particle is approximately 23 times its own weight.

Consider the triangular groove pattern of Fig.23. If the plates are covered with a thick corrosion - resistant coating, such as chrome, then an effective air-gap exists at P . In the prototype Jones separator of Fig.21 this gap is about 0.38mm, the pitch of the groove is 3.2mm and the perpendicular distance from the apex (Q) of the groove to the adjacent flat surface is 1.97mm. Approximate values of the local field at the points P and Q are 2.38T and 0.81T respectively, assuming a mean value for this plate system of 1.6T, as before. With a PQ of length 2.27mm, it may be assumed that a linear field gradient of 0.69 x $10^3 Tm^{-1}$ exists in the direction PQ of Fig. 23. Therefore, at P, the tractive magnetic force on the 52 mesh siderite particle is (eqn.(27)) 190 times the particle weight. It is also possible to calculate the normal and sideways pull on particle in the triangular groove plate pattern, by a line dipole approach in which the triangular groove pattern is 'imaged' in the flat surface and a calculation similar to that for the salient groove pattern is carried out.

Fig.22. View of pole and part of rotor of
Boxmag-Rapid HW436 Separator

In addition, Jones calculated typical Reynolds' number values for the flow values in grooved plate systems. For typical slurry flow velocities of around $0.5ms^{-1}$, Reynolds' numbers in the range 400→900 are obtained, the exact value being dependent on the details of the plate geometry and on the nature of the slurry. On the whole, therefore, in Jones-type devices, the flow between the plates is laminar, although some considerable turbulence is found at the entry point of the feed. The turbulence becomes minimal after about 0.08mm from the point of entry. This turbulence is of some advantage in that it helps to prevent the build-up of a thick deposit of particles at the point of entry and also

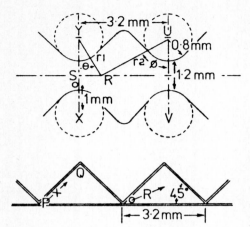

Fig.23. Geometry of salient and
triangular grooves
(after Jones)

to steer magnetic particles away from neutral (trojan) points in the field pattern (e.g.the point 7 of Fig.23) where the magnetic tractive force is at a minimum.

4.4 The electromagnetic filter

This is the popular name (ref.38) generally given to a highly successful commercial (Appendix 2) filter employing a random steel ball matrix in a water-cooled solenoid, as shown schematically in Fig.24. This filter is widely used in the nuclear power industry for the extraction of insoluble radio-active corrosion products (primarily sub-micronic magnetite) from the water circuits of PWR/BWR reactor systems. Prototype reduced scale forms of this filter have operated (ref.39) successfully in reactor loops with water velocities of up to $0.55ms^{-1}$. The prototype filter consists of an austenitic steel pipe, the inside diameter of which is 74mm, with a wall thickness of 7.5mm. The steel ball matrix length is 0.53mm, comprising 11kg of 13% chromium ferritic steel balls (M_s=0.3T) of 6mm diameter. The solenoid field value is around 0.2T.

4.5 Laboratory HGMS devices

A variety of small, relatively simple, HGMS laboratory devices have been developed and marketed commercially. These are used widely in the evaluation of the suitability of small samples for treatment by magnetic separation.

(a) Davis Tube

This is a laboratory concentrator frequently used as a rapid method for determing the magnetite content of ores (ref.25). The apparatus (Fig.25) consists essentially of a powerful electromagnet with pointed poles together with a long glass tube having facilities for intro-ducing into it the mineral sample as well as wash water. The device is sometimes fitted with a small motor for the purpose of agitating the glass tube. Magnetic material in the mineral sample is held between the poles and the gangue is washed down the tube.

TABLE 6
Throughput capacities (in th^{-1}) of double rotor (DP) and single rotor (P) Jones separators for Brazilian haematite(courtesy of KHD).

Type	Weight kg approx.	Throughput capacity(th^{-1})
DP 335	114000	for approx. 180
DP 317	98000	for approx. 120
DP 250	70000	for approx. 75
DP 180	41700	for approx. 40
DP 140	29200	for approx. 25
DP 112	22400	for approx. 15
DP 90	16200	for approx. 10
DP 71	13400	for approx. 5
DP 71	9200	for approx. 2.5
P 40		for approx. 0.5

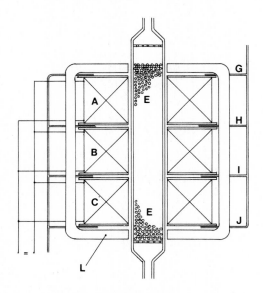

A B C − magnet coil winding
E − ball matrix
G H I J − water cooling
L − iron return circuit

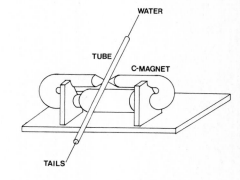

Fig.24. The electromagnetic filter
(after Dolle et al(ref.39)

Fig.25.The Davis tube

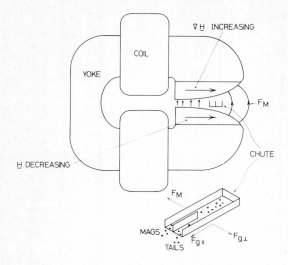

Fig.26. The Frantz Isodynamic
Separator (courtesy of
S.G.Frantz Co, Inc)

(b) Frantz-Isodynamic
Separator

This is a more complex laboratory device consisting of a
powerful horseshoe electromagnet with curved pole pieces
designed so that the field
gradient and the resultant
magnetic tractive force on a
particle are constant in the
operating zone (Fig.26).
Consequently, the influence
on particle movement is
uniform throughout this zone.
The proximity of a particle to
one or other of the pole pieces
is of no consequence as far as
its ultimate separation point
is concerned. Material travels down a chute between the pole pieces;
the chute is inclined to the horizontal both along its length and also
perpendicular to its length, the latter tilt ensuring that a component
of the gravitational force acts in direct opposition to the magnetic
force. The chute is normally agitated by a motor in order to help the
small particles of the powdered feed overcome the effects of friction.
Particles travelling down the sloping chute are subjected, in the
separating zone, to the action of a regulated isodynamic magnetic
force and are accelerated laterally in a fashion dependent   only on
their mass susceptibilities. By careful operation of this device it
is possible to separate very weakly susceptible powdered paramagnetic
materials. Final separation and collection of two particle species
is effected by a dividing knife edge at the bottom of the chute (Fig. 26)
The isodynamic nature of the magnetic force means that the dipolar
force, F_m, is constant across the width of the chute, where

$$F_m = \mu_o m (\chi_p / \rho_p) H \frac{\partial H}{\partial x}$$

and ρ_p is the particle density. If θ is the sideways tilt of the
chute from the horizontal then the lateral component of gravity, F_{GL},
is

$$F_{GL} = mg \sin\theta$$

so that the net lateral acceleration of the particle, that is

$$\frac{F_M - F_{GL}}{m}$$

$$= \mu_0 (\chi/\rho_p) H \frac{\partial H}{\partial x} - g\sin\theta,$$

is independent of particle size. The device therefore produces a separation based only on the mass susceptibility of the particle, a feature which makes it truly unique among separation devices. Typical operating values of this device are summarised in ref.40. . In practice θ is typically of the order of a few degrees. The device may also be operated in the so-called free-fall mode in which the pole gap of Fig.26 is rotated to the vertical position and the powdered particles are dropped under gravity through the gap. In this mode it is well established (ref.5,40) that it is possible to separate even diamagnetic powders. Fig.27 shows the results achieved by Kester and co-workers (ref.41) on the removal of inorganic pyritic sulphur from various dry pulverized coals using this device.

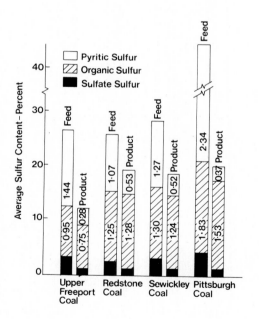

Fig.27. Results of separation of inorganic sulphur from dry powdered coals using the Frantz Isodynamic Separator (after Kester et al.ref.41)

The growth of the technological society has created social and economic problems of rapid depletion of material resources and of increased levels of environmental pollution. The need has grown, increasingly, for the exploitation, through magnetic separation and other competing technologies, of mineral deposits previously regarded as uneconomic. Similarly, the tightening of environmental control has stimulated investigation of a range of problems in the field of magnetic separation beyond the capabilities of any of the aforementioned conventional separation devices. This has led to the development of a new generation of HGMS devices and there follows a description, in some detail, of the designs, modes of operation, fundamental principles and areas of application of these devices.

4.6 Filamentary matrix separation systems

Since about 1970 (ref.42) a new generation of HGMS devices has evolved, for reasons of the kind described above, with the capability of removing efficiently, from wet or dry fluid streams, weakly paramagnetic particles of diameters of the order of one micron. An outstandingly successful device of this type, first introduced by Kolm (ref.43) contains in the separating zone, a highly porous matrix of filamentary ferromagnetic material, such as a ferritic stainless steel wool, placed in a strong uniform magnetic field (Fig.28). The effectiveness of this type of device has already been demonstrated through eqn.(8) in which it was shown that a capturing structure within the field volume of a separation device must be matched as closely as is practicable, to the sizes of individual particles.

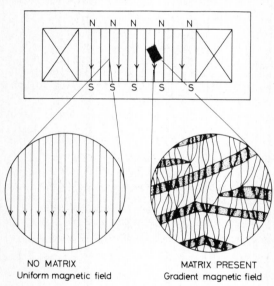

NO MATRIX
Uniform magnetic field

MATRIX PRESENT
Gradient magnetic field

Fig.28. Schematic representation of the field gradients produced in a Kolm-type HGMS system.

Fig.29 shows electron micrographs of three quite distinct types of filamentary matrix material. Photograph 29(a) shows a single strand of 12μm diameter, 304 grade stainless steel sliver (Bekaert) which has been used in the construction of self-supporting parallel bundles used in axial filters(ref.45).

It has already been shown (Fig.8) that, for particles in the micron size range in liquid carrier streams, the competing hydrodynamic drag reduces more slowly with decreasing particle size than the magnetic tractive force. It follows, therefore, that an efficient magnetic separation device for colloidal particles requires a much larger value of the magnetic field gradient than that of any of the devices already discussed.

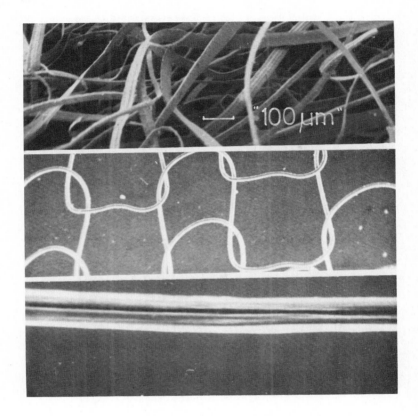

Fig.29   Some non-corrosive ferromagnetic fibres used in
         filamentary matrices, including (a) 430 grade
         stainless steel wool, (b) a knitted sheet of
         continuous 430 grade 50μm diameter wire and (c)
         a cylindrical strand of 304 grade 12μm diameter
         sliver (see Table 7)

Consider a random filamentary matrix consisting of small ferromagnetic fibres, of cylindrical cross-section, placed in a high
uniform background magnetic field. Obviously, in such a structure,
an appreciable fraction (refs.46,47) of the fibres are oriented with
their axes approximately orthogonal to the field. Consequently, these
fibres are magnetized, transversely, to saturation and produce
gradients of the order of M_s/a (see Eqn.(A9)¹ in Appendix 1) in the
immediate vicinity of appreciable sectors of the fibre surface.

TABLE 7

Some Manufacturers and Suppliers of
Magnetic Fibres/Steel Wool

Company	Activities
Brunswick Corporation 45 Woodmont Roa Milford Conn.06460 U.S.A.	Stainless steel silver
International Steel Wool Co. 1800 Commerce Road Springfield Ohio 45501 U.S.A. Suppliers:All Stainless Inc. 75 Research Road Hingham. MA.02043 U.S.A.	Stainless steel wool
Indiana General Magnet Products Valparaiso. In.46383 U.S.A.	CuNiFe fibre in 300mm lengths
N V Bekaert S.A. 8550 Zwevegem-Belgium	430 grade stainless steel wool; fine stainless steel slivers & metal fibre web
Temco Ltd. Stowfield Lydbrook Glos.GL17 9NH. U.K.	Continuous wire bobbins of radiometal, nichrome.
Goodfellow Metals Ltd. Cambridge Science Park Mitton Road Cambridge CB4 4DJ. U.K.	Nickel wire
The Expanded Metal Co.Ltd. P O Box 14 Hartlepool TS25 1PR.U.K.	Expanded metal mesh in 430 grade stainless steel
Gilby-Brunton Ltd Seamill Musselburgh Scotland, U.K.	Continuous wire bobbin's of stainless steel

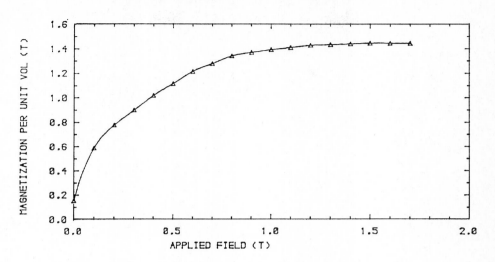

Fig.30. Transverse magnetisation curve for 50μm
        diameter 430 grade stainless steel wire
        (courtesy of S.D.Berger)

Field gradients of the order of $10^5 Tm^{-1}$ are not uncommon in commercial devices employing this type of filter structure. The applied field required to saturate the fibres depends on their ferromagnetic composition and on the detailed geometry of the fibre cross-section. Fig.30 shows a transverse magnetisation curve for the cylindrical fibres of ferritic stainless steel shown in Fig.29(b). Clearly, because of the large transverse demagnetizing factors for the cylindrical geometry (ref.18) these filamentary matrices are difficult structures to magnetize to saturation levels. Nesset and Finch (ref.48) have shown that the magnetisation, M, of fibres in this transverse orientation can be related to the saturation magnetisation M_{SB} of the bulk medium through the formula of Clarkson (ref.49).

$$M = M_{SB} \exp \ (-K/(H_o - N_D M)$$

where N_D is the demagnetizing factor of the fibre and where K is a constant. Magnetisation curves for expanded mesh stainless steel (ref.48) and for nickel fibres (refs.33,48) can be found in the literature. There exists a wide variety of non-corrosive ductile

magnetic alloys with suitable properties for the manufacture of
matrix material. Probert (ref.50) has measured the transverse
magnetization of fibres of various materials supplied by the authors,
including permalloy, radiometal and permendur, some suppliers of
which are listed in Table 7.

High values of applied magnetic field are required to produce high
field gradients along the fibre surfaces. Conversely, when this field
is turned off, the residual magnetization of the fibres is very low.
Consequently, even strongly magnetic particles can easily be removed
by backwashing when the applied field is reduced to zero.

Efficient filamentary HGMS devices are crucially dependent on the
economic production of high magnetic fields on a large scale. Much
of the credit for the commercial and scientific success of filamentary
HGMS must be attributed to Marston (ref.51) for significant design
improvements to the magnetic circuit of the large-scale iron-bound
solenoid. This design, first marketed by Sala (Appendix 2)
is perhaps seen to best advantage by reference to the diagrammatic
comparisons first made by Mitchell et.al (ref, 8) and reconstructed
in Fig.31. Here the coil windings, field working volume and magnetic
return circuit are superimposed on the magnetic circuit of a
conventional Weiss-type (ref.18) electromagnet of identical field
volume. In the novel Sala design, the water-cooled coil windings are
immediately adjacent to the working volume and these are totally
surrounded by a highly-efficient iron return circuit of relatively;
modest volume. This Marston-type magnet is capable of producing
background fields in the range 1 to 2 T at comparatively modest
inputs of electrical energy and at virtually any desired scale of
operation.

The typical mode of operation of a cyclic Kolm-Marston type of
separation device can be understood easily from the schematic diagrams
of Fig.32. In the left hand diagram the mixed product, in the form
of a wet suspension or slurry, is shown as being pumped vertically
upwards through the filamentary matrix with the magnetic field
switched on. The magnetic particles in the mixed product are trapped
at the surface of the magnetized fibres, while the non-magnetic
particles and carrier fluid pass easily through the canister. The
matrix, which, typically, has a porosity of the order of 0.95,
presents an extremely low flow resistance of the order of $10^6 \mathrm{Nsm}^{-4}$

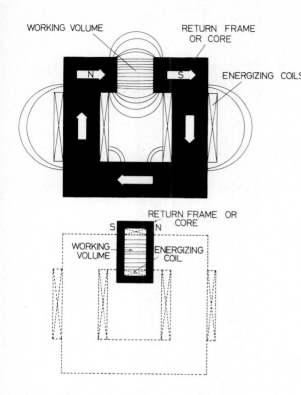

Fig.31 Schematic representation of the flux return circuits of a conventional (upper diagram) and a Marston-type electromagnet (after Mitchell et al. (ref.8))

to the wet feed (ref.52). When the matrix has become loaded to saturation with magnetic particles the flow is halted, as shown in the right-hand diagram. The magnetic field is reduced to zero and the magnetic particles easily washed from the matrix via a separate flushing system, as shown. This type of so-called 'cyclic' (refs.8,9) HGMS device is used, typically, to process wet feed streams containing a relatively modest level of magnetic contamination. Such devices are extremely useful in a variety of large-scale water treatment and wet mineral processing applications (see Table 8.) Fig.33 shows one such large installation, manufactured by P.E.M. (Appendix 2), used in the beneficiation of kaolin at English Clays Lovering Pochin and Co.Ltd, St Austell, U.K. The device shown has a 3.1 m bore with a background magnetic field of around 2.0T.

In applications where a valuable magnetic component, requiring separation, exists as a sizeable component of a feed stream, rotary filamentary separation systems known as Carousel systems (ref.9) have been introduced of somewhat similar construction to the afore mentioned rotary Jones-type separators. Fig.34 shows a schematic design of one such system marketed by Sala for applications such as high-speed recovery of colloidal magnetite in dense media systems.

5. THEORY OF PARTICLE CAPTURE IN FILAMENTARY MATRICES
5.1 Calculation of particle trajectories

218

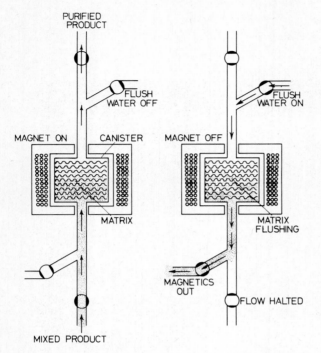

PURIFIED
PRODUCT

FLUSH
WATER OFF

FLUSH
WATER ON

MAGNET ON   CANISTER   MAGNET OFF

MATRIX

MATRIX
FLUSHING

MAGNETICS
OUT

FLOW HALTED

MIXED PRODUCT

Fig.32 Fluid flow in a cyclic Kolm-Marston
type of HGMS system.

A completely general
treatment of the magneto-
static and of the hydro-
dynamic behaviour of a
microscopic particle
moving in (say) a
totally random fila-
mentary structure of
ferromagnetic fibres
in the presence of a
steady magnetic field is,
clearly, a theoretical
problem of considerable
complexity.
The problem can be
simplified considerably,
however, by considering
in the first instance the
interaction between a
single particle and a
single fibre, with the
further simplification

TABLE 8
Technical characteristics of the Series LP range of cyclic, automatic backflushing filters marketed by Sala Magnetics Inc.

Model	Field (T)	Matrix Area (ft^2)	Feed throughout range (gpm)	Power input(KW)
305-15-5	.5	77	5900-19650	75
214-15-5	.5	37.8	2900-9670	55.5
152-15-5	.5	19.2	1540-5130	42.5
107-15-5	.5	9.3	720-2400	35.5
76-15-5	.5	4.7	360-1200	28
56-15-5	.5	2	200-670	20
38-15-5	.5	0.83	190-300	18
10-15-5	.5	0.06	5-18	9.4

Fig.33   A P.E.M. High Intensity Magnetic
              Separator used by English Clays
              Lovering Pochin and Co.Ltd. for the
              production of ceramic china clays

that the fibre is always orthogonal to the external magnetic field and
with the particle moving, initially, in any one of three possible
principal directions (shown in Fig.35) allowed by symmetry in relation
to the fibre/field combination.   When the general fluid flow
direction and, consequently, the initial particle direction and the
external magnetic field are parallel or anti-parallel to one another
(Fig.35(a)),   this is known as (ref.47) the longitudinal (L-)
configuration. When the initial fluid flow direction and the field
are orthogonal to one another (Fig.35(b)) this is termed the transverse
(T-) configuration.   The third alternative configuration, when the
initial fluid flow direction is parallel to the fibre axis, is known

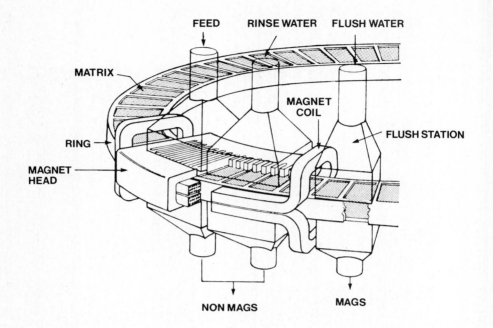

FEED   RINSE WATER   FLUSH WATER

MATRIX

MAGNET
COIL

RING

FLUSH STATION

MAGNET
HEAD

NON MAGS   MAGS

Fig. 34   Carousel continuous large-scale
             HGMS system used in both wet and
             dry mineral separation applications
             (Courtesy of Sala Magnetics Inc.)

as the axial configuration (refs 45,53).

The more general situation of particle-fibre interaction involving
magnetic and hydrodynamic interactions with multiple fibre arrays, as
well as the problem of arbitrary orientation of the fibre in relation
to the initial particle direction are dealt with in §5.5 as
extensions of the theoretical treatment of the interactions relating
to the three arrangements of Fig.35.

Consider a single constituent cylindrical fibre of the matrix
(fig.36).   This fibre, of radius a   and of saturation magnetization
M_s, is magnetized fully by a strong external magnetic field, H_o.
A fluid (which may be regarded here as incompressible and frictionless)
flows around the fibre under conditions of potential flow. At large
distances (>>a) from the fibre, the fluid flows with uniform speed in
a direction making an angle γ with the x -axis of Fig.36. If the
fluid contains colloidal paramagnetic (or diamagnetic) particles then

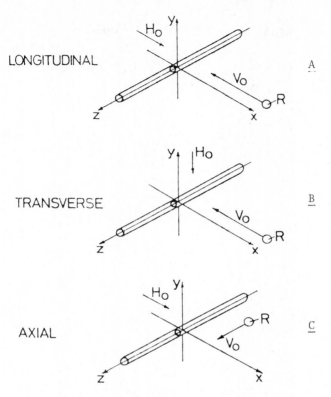

Fig.35. The (a) longitudinal (b) transverse and (c) axial
capture configurations (ref.44)

these will be attracted towards particular parts of the surface of
the fibre and some may be held at that surface. The equations of
motion of such particles are now derived in terms of the cylindrical
polar coordinates (r, θ) of Fig.36, the z-axis being coincident with
the axis of the cylindrical fibre.

   If, for the reasons stated earlier, gravitational forces are
ignored, then the dynamic behaviour of the particle may be described
by an equation of the form

$$m \frac{d\underline{v}}{dt} = \underline{F}_m - \underline{F}_d$$

(28)

where F_d is the hydrodynamic drag force produced by the differential
velocity, $\underline{V}$, of the particle relative to the fluid. The magnetic force
$\underline{F}_m$, acting on the particle is given, in general terms by Equn.(4).

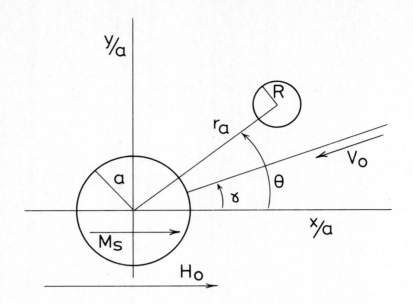

Fig.36. Coordinate system for particle-wire interaction

However, to find its precise form in this particular geometry, it is necessary to find a satisfactory solution of Laplace's Equation for the boundary conditions of Fig.36. This solution(using Kennelly SI units) is shown in Appendix 1. F_M of eqn.(4) may now be expressed in terms of eqns A(14) and A(15) as radial and azimuthal forces respectively of the form

$$F_{M,r} = -\frac{1}{2}\mu_o(\chi_p-\chi_m)V_p \ (4A/r^3)(H_o\cos 2\theta+A/r^2) \quad (a) \tag{29}$$

$$F_{M,\theta} = -\frac{1}{2}\mu_o(\chi_p-\chi_m)V_p \quad (4H_o A/r^3)\sin 2\theta \quad (b)$$

where $A = M_s a^2/2\mu_o$ $\qquad\qquad\qquad\qquad\qquad\qquad$ (30)

If the net volume susceptibility $\chi=(\chi_p-\chi_m)$ in eqns (29) is positive then the radial component of the dipolar force between the particle and the fibre is attractive, provided that $F_{M,r}$ is negative. This is the case for paramagnetics, provided

$$A/r^2 + H_o\cos 2\theta > 0 \tag{31}$$

The component $F_{M,r}$ becomes repulsive, for paramagnetics, in two symmetrical curved spatial regions adjoining the fibre surface, the boundaries of these regions being determined by a variable critical angle θ_c, where

$$\theta_c = \text{arc tan} \quad (1+K/r_a^2)/(1-K/r_a^2) \tag{32}$$

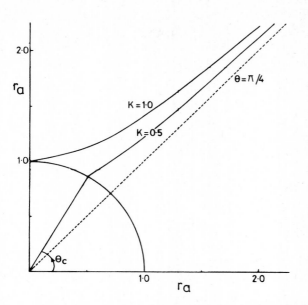

Fig.37. Attractive and repulsive regions of space adjacent to a cylindrical fibre, when H_o is in the θ = 0 direction.

Clearly, θ_c, which is strongly dependent on the parameter, K, (Fig.37), determines the respective sectors of the fibre surface which are attractive (or repulsive to paramagnetic or to diamagnetic particles.

From eqn.(12) it can be seen that $\underline{F}_D$ = $6\pi\eta VR$. If potential flow is assumed for the passage of the fluid around the fibre, the set of cylindrical harmonics used in Appendix 1 to generate the magnetostatic potential may be used to calculate the velocity. The radial and azimuthal components of the hydrodynamic drag force $\underline{F}_D$ are shown (see Appendix 1) to be, respectively

$$F_{D,r} = 6\pi\eta R \left[\frac{dr}{dt} - V_o\cos(\theta-\gamma)(1-a^2/r^2) \right] \quad \text{(a)} \tag{33}$$

$$F_{D,\theta} = 6\pi\eta R \left[\frac{d\theta}{dt} + V_o\sin(\theta-\gamma)(1+a^2/r^2) \right] \quad \text{(b)}$$

where the fluid is moving, at large distances from the fibre, with speed V_o in a direction at an angle γ to the x-axis of Fig.36. It can easily be shown (ref.47) that the radial and azimuthal components of the inertial term on the left hand side of eqn.(28) may be written as

$$F_{i,r} = (4/3)\pi R^3 \rho_p \left\{ \frac{d^2r}{dt^2} - r\left(\frac{d\theta}{dt}\right)^2 \right\} \quad \text{(a)}$$

$$\tag{34}$$

$$F_{i,\theta} = (4/3)\pi R^3 \rho_p \left\{ r\frac{d^2\theta}{dt^2} + 2\frac{dr}{dt}\left(\frac{d\theta}{dt}\right) \right\} \quad \text{(b)}$$

respectively, where ρ_p is the particle density. Eqn.(28) can therefore be re-expressed in the form

$$(2\rho_p R^2/9\eta) \{ \frac{d^2 r_a}{dt^2} - r_a (\frac{d\theta}{dt})^2 \} + \frac{dr_a}{dt} \quad =$$

$$V_0/a)(1-1/r_a^2)\cos(\theta-\gamma) - (V_m/a)\{(M_s/2\mu_0 H_0)/r_a^2 + \cos 2\theta\}(1/r_a^3) \quad (35)$$

for radial motion, and

$$(2\rho_p R^2/9\eta)\{r_a(\frac{d\ \theta}{dt^2}) + 2(\frac{dr_a}{dt})(\frac{d\theta}{dt})\} + r_a \frac{d\theta}{dt} \tag{36}$$

$$= -(V_0/a)(1+1/r_a^2)\sin(\theta-\gamma) - (V_m/a)(\sin 2\theta/r_a^3)$$

for azimuthal motion, where $r_a = r/a$ and where $V_m = 2(\chi_p - \chi_m)H_0 M_s R^2/9\eta a$ is a dimensionless parameter termed the 'magnetic velocity' (refs. 46,47). It can be seen clearly for the case of colloidal particles in liquid carrier streams that $(2\rho_p R^2/9\eta)$ is exceedingly small, typically of order 10^{-6}. As such, the inertial term on the left-hand side of both eqns.(35) and (36) may be ignored, in which case these equations together describe the loci of equilibrium states of the particle brought about by the balance of the hydrodynamic and the magnetic forces. Eqns.(35) and (36) then reduce to the simplified inertialess forms

$$\frac{dr_a}{dt} = (V_0/a)(1-1/r_a^2)\cos(\theta-\gamma) - (V_m/a)\{K/r_a^5 + \cos 2\theta/r_a^3\} \quad (36)$$

$$r_a\frac{d\theta}{dt} = -(V_0/a)(1+1/r_a^2)\sin(\theta-\gamma) - (V_m/a)(\sin 2\theta/r_a^3)$$

where K (eqn.(9)) is a dimensionless parameter (ref.45) whose physical influence is examined in detail later. When $\gamma = 0$, eqn.(38) describes the trajectories of particles in the so-called longitudinal configuration Fig.35(a). When $\gamma = \pi/2$, they describe motion in the transverse configuration. If eqn.(38) is divided by eqn.(39) the time-independent equation

$$\frac{1}{r_a}\frac{dr_a}{d\theta} = \frac{(1-1/r_a^2)\cos(\theta\gamma) - (V_m/V_0)\{K/r_a^5 + \cos 2\theta/r_a^3\}}{-(1+1/r_a^2)\sin(\theta-\gamma) - (V_m/V_0)(\sin 2\theta/r_a^3)} \tag{40}$$

is obtained, which describes the particle trajectories. Clearly, the

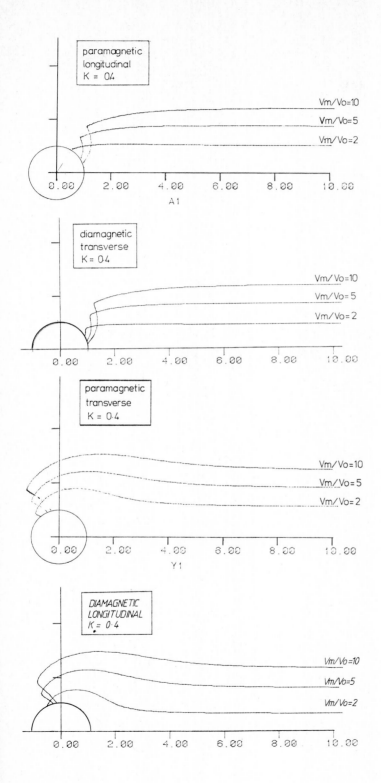

Fig.38.
Limiting tra-
jectories for
paramagnetic
and for dia-
magnetic
particles in
both the L-
and T-con-
figurations,
with

$\dfrac{V_m}{V_o}$ values

of 2,5 and 10.
(ref.20)

solution of this particular equation is dependent upon the
dimensionless parameter (V_m/V_o) and not on the values of V_m and V_o
separately. The equation can be solved numerically, with ease, by
employing a standard procedure such as a fourth order Runge-Kutt
technique (Appendix 3) using as simple a device as a pocket calculator.
In this way, the trajectories of particles in the region of a magnetised fibre can be calculated for various initial coordinates (r_{ai}, θ_i)
of the particle. A particle may be regarded, in the first instance,
as captured if its trajectory terminates at the fibre surface $(r_a=1)$
and not captured if its trajectory passes well beyond the fibre.
Figs.38 shows numerically evaluated particle trajectories for both
paramagnetic and diamagnetic particles in the L $(\gamma=o)$ and $T(\gamma=\pi/2)$
configurations. All the trajectories shown are for starting positions
$x_a (=r_{ai}\cos\theta_i)$ at which the influence of the fibre is negligible. The
limiting initial value of y_a $(=r_{ai}\sin\theta_i)$ for which capture occurs
(Fig.38) is defined as the capture radius, $r_c (=r_{ca}a)$, of the fibre.
From eqn.(40) it can be seen that r_{ca} is a function both of (V_m/V_o) and
of K. Figs.39 and 40 show computed (ref.20) values of the capture
radius as a function of the parameter, K, for a fixed value of $V_o (=5)$,
for both paramagnetic and diamagnetic particles in the L- and T-
configurations respectively.

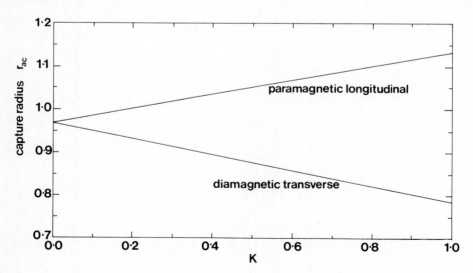

Fig.39. Dependence of normalized capture radius (r_{ca}) on K for
paramagnetic and diamagnetic particles in the L- and
T-configurations respectively $(V_m/V_o=5)$.

A useful approximate analytical solution of eqn.(40), for the case where $V_m/V_o \gtrsim 10$ and where K is reduced to zero, is obtainable easily (ref.56) for either of the principal orientations ($\gamma = 0, \pi/2$). Here, it is a straightforward matter to show, for the case $\gamma = 0, K = 0$, that the initial (normalized) vertical offset distance, y_{ai}, of the particle from the x-axis of Fig.38 (where $y_{ai} = r_{ai} \sin \theta_i$) can be related to other points on the particle trajectory via the equation

$$y_{ai} = r_a(1 - \frac{1}{r_a^2})\sin\theta + (\frac{V_m}{2V_o}) \frac{\sin 2\theta}{r_a^2} \qquad (41)$$

with a similar result for the case $K = 0$, $\gamma = \pi/2$. Eqn.(41) can be transformed into a reduced cubic equation in r_a (ref.54) for which analytical solutions exist. The nature of these solutions is dependent on the sign of the discriminant, a singular result being obtained for the case where the discriminant is identically zero. In this case, two equal real roots and a third zero root are found, the real roots referring to the singular point (r_{as}, θ_s) (refs.33,35,55) that is the turning point on the trajectory, (Fig.39) with the limiting initial offset distance y_{ai} at which the particle is just captured. Fig.41

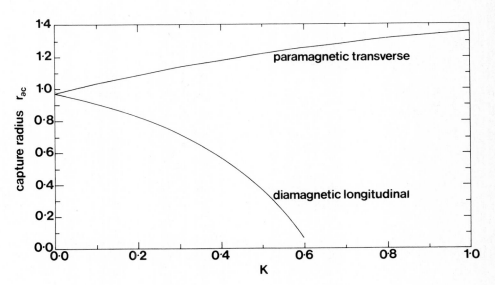

Fig.40. Dependence of normalized capture radius (r_{ca}) on K for paramagnetic and diamagnetic in the T- and L-configurations respectively ($V_m/V_o = 5$).

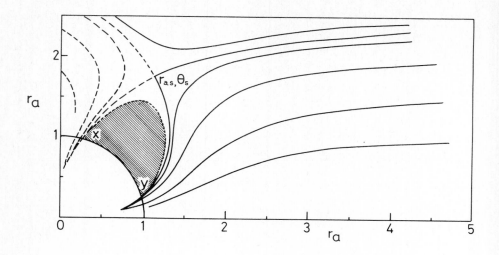

Fig.41. Longitudinal, transverse and "forbidden" trajectories
generated from the reduced cubic equation (c.f. eqn.(41)) for the
case $\frac{V_m}{V_o}$ = 10. Note the existence of the singular point (r_{as}, θ_s).

shows a family of particle trajectories, generated (ref.20) from the
above-mentioned reduced cubic equation. Here, the dashed curves may
be interpreted physically as final portions of capture trajectories
in the T-configuration. The dotted curve is a solution of the
reduced cubic describing the motion of a paramagnetic particle
which finds itself placed artificially in a 'forbidden' region of
space around the fibre. The hatched region of space between the
critical L- and T-trajectories and the fibre is, under normal circumstances, inaccessible to any particle(with $\frac{V_m}{V_o}$ = 10 in this particular
case) moving towards the fibre from anywhere on the right-hand side
of Fig.41. If, however. a paramagnetic particle is placed, arti‑
ficially, at the point X on the surface of the fibre it will be
repelled, initially, from the surface of the fibre and move along
the dotted arc shown in Fig.41 until it arrives at the point Y on
the surface where the radial magnetic force is attractive. The
reverse holds for a diamagnetic particle.

Uchiyama (ref.35) has refined the simple analytical solution of
eqn.(41) to include the short-range force term of eqn.(29). His

solutions are only approximate, but show a fairly impressive correlation with the numerical predictions of the Runge-Kutta computations. His analytical solutions for both L- and T-configurations for the case of ideal flow are summarized in Table 9. His analysis may, in fact, be generalized to include diamagnetic capture by changing K to a negative quantity (ref.14). In this Table and from a study of

TABLE 9

Approximate analytical expressions for r_{ca} for paramagnetic particles in the L- and T-configurations (ref.35)

Field/Flow	Ideal flow	V_m/V_o	Normalized Capture Radius
Longitudinal	Ideal	$> \dfrac{2}{1-K}^{\frac{1}{2}}$	$\dfrac{3\sqrt{3}}{4}\left(\dfrac{V_m}{V_o}\right)^{1/3} \left(\dfrac{V_m}{V_o} - \dfrac{\sqrt{3}}{2}-0.052K\right)^{1/3}$
		>3	
		$\gtrsim 0.1$	$\dfrac{V_m}{2V_o}\left[\dfrac{\left(1-K^2\right)^{\frac{1}{2}}+K(\pi-\cos^{-1}K)}{1+\dfrac{4K}{5}\dfrac{V_m}{V_o}}\right]$
Transverse	Ideal	$\gtrsim 3$	$1.30\left(\dfrac{V_m}{V_o}\right)^{1/3}\left(\dfrac{V_m}{V_o} - 0.87-0.36K\right)^{1/3}$
		$\gtrsim 0.1$	$\dfrac{V_m}{2V_o}\left[\dfrac{(1-K^2)^{\frac{1}{2}}+K(2\pi-\cos^{-1}K)}{1+\dfrac{4K}{5}\dfrac{V_m}{V_o}}\right]$

the literature (refs. 20,35,56) it can be deduced that the expressions for r_{ca} fall into two distinct regimes. This can best be seen from the L-configuration in which it is clear that, for paramagnetic particles, the angle of arrival on the surface of the fibre increases as V_m/V_o decreases (Fig.38). This process continues until the critical angle, θ_c , is reached (eqn.(32)) whereupon $F_{M,r}$ changes sign. In ideal flow conditions, a small particle is stable on the surface of the fibre in this configuration if $\dfrac{d\theta}{dt} < 0$, that is (eqn.(39) with $r_a = 1, \theta = \theta_c$)

$$\cos\theta_c = \left| \dfrac{V_m}{V_o} \right|^{-1}$$

but, from eqn.(32),

$$\cos\theta_c = \left(\dfrac{1-K}{2}\right)^{\frac{1}{2}}$$

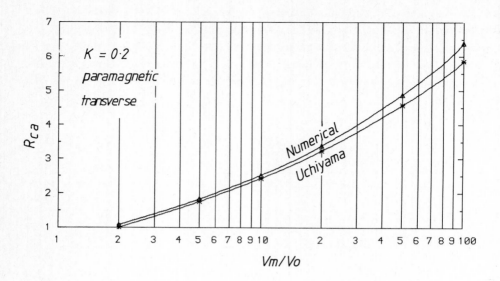

Fig.42. Variation of normalized capture radius, R_{ca}, with $\frac{V_m}{V_o}$ for paramagnetic particles (ref.20).

In other words, a singular point exists outside the fibre provided $|V_m/V_o| > (2/(1-K)^{\frac{1}{2}}$, (Table 9). For values of (V_m/V_o) of this order, or greater, r_{ca} is found to exhibit a 1/3 power dependence on $|V_m|V_o|$ (Fig.(42), although numerical analyses (refs.20,57) tend to favour a 3/8 power dependence. Friedlaender et al. (ref.57) arrive at an expression of the form

$$r_{ca} = 1.2 \left| \frac{V_m}{V_o} \right|^{\frac{3}{8}} - \frac{(1-K)}{5} \left| \frac{V_m}{V_o} \right|^{\frac{1}{2}} \tag{42}$$

for the T-configuration, while M.K.Wong (ref.20) arrives at the expression

$$r_{ca} = \left| \frac{V_m}{V_o} \right|^{\frac{3}{8}} + \frac{K}{5} \left| \frac{V_m}{V_o} \right|^{\frac{2}{5}} \tag{43}$$

for the configuration. For low values of $|V_m/V_o|$, of order 0.1, both analytical and empirical studies favour a linear relationship between r_{ca} and $|V_m/V_o|$ (Fig.43)

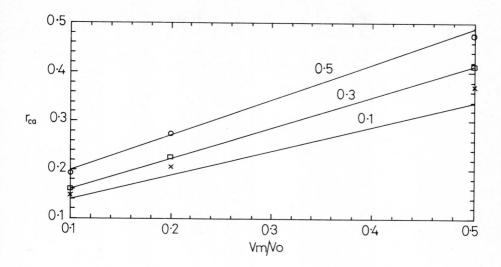

Fig.43. Variation of normalized capture radius, r_{ca}, for paramagnetic particles in the L-configuration for low values of Vm/Vo.(ref.20)

A variety of published calculations (refs.33-37) of particle capture in filamentary matrices have extended the above-described single wire theory to include the combined field and flow effects of a regular array of wires. The most comprehensive treatment of this problem is that of Hayashi and Uchiyama (ref.37) who have recently published particle trajectories and capture efficiencies for the case of two capturing layers of regularly arranged wires (Fig.44). In this calculation, they use the Isenstein (ref.34) expression for the magnetic potential, $\emptyset$, at the normalized coordinates x , (Fig.44) of the form

$$\emptyset = -x_o H_o + \frac{M_s}{2\mu_o} \sum_{m=-\infty}^{+\infty} \frac{x_a}{x_a^2 + (y_a - m\ell/a)^2}$$

$$+ \frac{M_s}{2\mu_o} \sum_{m=-\infty}^{+\infty} \frac{x_a - X_a}{(x_a - X_a)^2 + (y_a - Y_a - m\ell/a)^2}$$

(44)

$$= -H_o \left\{ x_a - \tfrac{1}{2}K\beta \left[\frac{\sinh\beta x_a}{\cosh\beta x_a - \cos\beta Y_a} \right] \right.$$

$$\left. -\tfrac{1}{2}K\beta \left[\frac{\sinh\beta(x_a - X_a)}{\cosh\beta(x_a - X_a) - \cos\beta(Y_a - Y_a)} \right] \right\} , \tag{45}$$

where $\beta = 2\pi a/\ell$, $x_a = x/a$, $y = y/a$, $X_a = X/a$, $Y_a = Y/a$. Also $K = M_s/2\mu_o H_o$ is the inter-wire spacing in each layer, X is the inter-layer separation and Y is the lateral offset of the wires of the second layer relative to the first (Fig.44).

The fields and dipolar forces generated by this wire array can easily be obtained using the procedures and equations of Appendix 1. It should be noted that the expression for fields and dipolar forces are very similar to those presented by the authors (ref.58) in their calculation of the particle capture by a periodic magnetic domain configuration. Hayashi and Uchiyama also use the complex potential stream function approach to describe the inviscid flow through wire arrangements of the type shown in Fig.44. Their results show clearly that, in the case of weak capture ($V_m/V_o < 1$), the capture efficiency of the double layer is essentially twice that of an individual layer. However, in the case of stronger capture, the performance is less than twice the value of an individual layer by an amount dependent on the extent to which geometrical masking of the second layer is present (Fig.44).

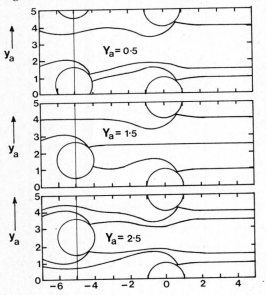

Fig.44.Capture characteristics for para-magnetic particles of a two-layer wire array in the L-configuration as a function of the normalized offset parameter, Y_a (ref.37).

If the situation is considered in which the carrier fluid is gaseous, the inertial terms of eqns.(35) and (36) become significant. Particle

trajectories are then only obtainable by numerical solution of
these two equations. Such numerical studies have been carried
out by Simons et al (refs.33,59) and, more recently, by
Walker (ref.36). The former have shown that the systematics of
capture can not be described satisfactorily here (by contrast
with the inertialess approximation) by the dimensionless parameters
V_m/V_o and K alone, but, instead, ignoring gravitational effects,
by three distinct dimensionless parameters which include, in
addition to K, the Stokes parameter $S=2R^2V_o/9a\eta$ and an
additional parameter $W = \mu_o^2H_o^2/\rho_pV_o^2$. It is interesting to
note that the product SWK is identical (but for a factor of 2),
to the parameter V_m/V_o.

Both studies (refs.36,59) have revealed that particle capture
trajectories are more complicated than their equivalent forms in
the inertialess limit. Fig.45 shows some limiting trajectories
computed by Walker for paramagnetic particles in the L-configuration. Fig.46 shows the dependence of r_{ca} on S for various
values of W. Two further points emerge from this diagram.
Firstly, r_{ca} is seen to be dependent solely on the parameter SWK
(that is, on V_m/V_o) for values of the Stokes number in the range
$0 < \log S < -0.17$. At higher Stokes numbers, inertial effects
become increasingly important and the parameter SWK increasingly
irrelevant. Secondly, at higher values of S and at low values
of W, r_{ca} becomes, in the inviscid approximation, essentially
independent of particle size and, as such, capture becomes isodynamic approximately.

5.2 The axial filter
(a) Ideal flow

The motion of a particle in the axial capture configuration
(Fig.35(c)) can be described conveniently (refs.44,45,60) in terms
of normalized cylindrical polar coordinates, r_a, θ, Z_a, where $Z_a = Z/a$.
Here, the fibre axis and the fluid flow are parallel to one another
and to the Z-axis. If the higher symmetry of this third configuration is considered and if inertial terms, once again, are
neglected, the radial and azimuthal equations of motion eqns.(38)
and (39)), reduce, here, to the simplified forms

234

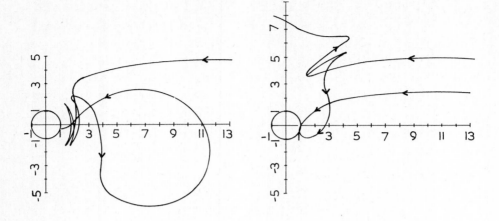

Fig.45. Particle trajectories for paramagnetic particles in the
L-configuration computed, using eqns.(35 and (36), for
the high inertia, low drag situation such as when the
carrier fluid is a gas. The left hand diagram shows
trajectories, one of which is critical for the case
K=1.0, S=10, W=10. The righthand diagram shows the
critical trajectory for the case K=1.0,S=100,W=10. (after
Walker ref.36).

$$\frac{dr_a}{dt} = -\left(\frac{Vm}{a}\right)\left(\frac{K}{r_a{}^5} + \frac{\cos 2\theta}{r_a{}^3}\right) \tag{46}$$

$$r_a\frac{d\theta}{a} = -\left(\frac{V_m}{a}\right)\left(\frac{\sin 2\theta}{r_a{}^3}\right) \tag{47}$$

respectively, with the motion of the particles along the length of
the fibre (in the inviscid approximation) described by a third
independent equation

$$\frac{dZ_a}{dt} = \frac{V_o}{a} \tag{48}$$

If time is eliminated from eqns.(46),(47) and(48), a pair of
differential equations

$$\frac{dr_a}{d\theta} = \frac{K}{r_a\sin 2\theta} + r_a\cot 2\theta \tag{49}$$

and

$$\frac{dZ_a}{d\theta} = \left(\frac{V_o}{V_m}\right)\frac{r_a}{\sin 2\theta} \tag{51}$$

allow an individual particle trajectory to be determined analytically. Eqn.(49) is a non -linear Bernoulli equation; eqn.(50 is
linear.   The solutions are

$$r_a{}^2 = -K\cos 2\theta + C\sin 2\theta \tag{51}$$

and
$$\bar{z}_a = \left[\frac{V_o}{V_M}\right] \left\{ \frac{K^2}{2} \log \frac{\tan\theta}{\tan\theta_i} - K(r_{ai}^2 - r_a^2) \right.$$

$$\left. - \frac{(K^2 + C^2)}{2} (\cos 2\theta_i - \cos 2\theta) \right\} \tag{52}$$

where C is a constant dependent on the initial (r_i, θ_i) of the particle in the x-y plane of Fig.36. Eqn.(51) describes a family of particle trajectories projected on to this x-y plane, some of which are shown in Fig.47, in the first quadrant. Similar trajectories can be arrived at, by symmetry, for the other three quadrants. Eqn.(51) is independent of the value and of the numrical sign of V_m/V_o and, therefore, is equally valid for paramagnetic and for diamagnetic particles. A paramagnetic particle starting at an angle, θ_i, close to the y-axis of Fig.36 circumscribes an orbit (with a maximum excursion radius r(max)) and arrives at the wire surface at an angle θ_f close to the x-axis. The reverse is true for a diamagnetic particle. For a paramagnetic particle, these angles are given by

$$\theta_{f,i} = \tfrac{1}{2} \arccos \left[\frac{-K \mp C(K^2 + C^2 - 1)^{\frac{1}{2}}}{K^2 + C^2} \right] \tag{53}$$

with the - and + signs corresponding to θ and θ , respectively. It can be shown that

$$r_a(\max) = (K^2 + C^2)^{\frac{1}{2}} \tag{54}$$

Eqn.(52) describes the progress of a particle along the z-axis, the nature of which is critically dependent on the initial position (r_{ai}, θ_i) of the particle. If $\theta_i = 0$, then the paramagnetic particle remains in the x-z plane throughout its trajectory, the latter being described analytically by the equation

$$z_a = \frac{V_o}{V_m} \left[\frac{(r_{ai}^4 - r_a^4)}{4} - \frac{K}{2}(r_{ai}^2 - r_a^2) + K^2 \ln\left\{ \frac{K + r_{ai}^2}{K + r_a^2} \right\} \right] \tag{55}$$

The authors have shown (ref.14) that diamagnetic particles originating in the y-z plane are similarly confined to that plane. It can be seen that the (normalized) wire length required to capture a particle can be obtained by evaluation of eqns.(52) and (55) at the end position $r_{ai} = 1, \theta = \theta_f$. For conditions of ideal flow it is possible to assign a given capture length to a wire for a wide variety of particles the times of capture of which, from various points of entry r_{ai}, θ_i in the $z = 0$ plane, are equal. The locus of all such points in the x-y plane at the front end of the wire ($z=o$), for particles of a given value of V_m/V_o, is given approximately by

$$r_{ai}^2 = K + 2 \left[1 + \frac{(K-1)^2}{8L_{aM}} \right] L_{aM}^{\frac{1}{2}} \cos\theta_i \quad , \qquad (56)$$

where

$$L_{aM} = \left(\frac{V_m}{V_o} \right) z_a . \qquad (57)$$

and is known as the isochronal curve (ref.45).

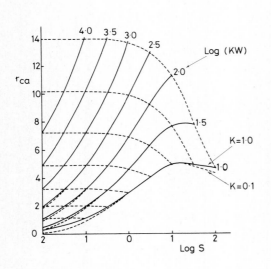

Fig.46. r_{ca} for paramagnetic particles in the L-configuration as a function of the logarithm of the Stokes number S. The dotted curves connect points of equal Vm/Vo (ie SWK) showing that r_{ca} is a function only if V_m/V_o up to log S = -0.18. Note that, at high values of S(10-100), r_{ca} becomes approximately constant. (After Walker ref.36).

(b) Laminar flow

The present authors and and co-workers(refs 17, 61,62) have extended the above described theory to include not only the effects of velocity laminations in the fluid surrounding the curve but also the effect of the adjacent wires within an axial filter, comprising a parallel bundle of identical wires, on the fluid flow pattern and on the capture process itself. In this calculation, the wire configuration considered is a two-dimensional lattice in which the parallel wires are ordered in a hexogonal structure. For the fluid flow, a 'muffin' tin velocity well is centred on each constituent wire within a hexagonal peri-meter. The velocity profile within the cell has approximately rotational symmetry around each wire, diverging from full radial

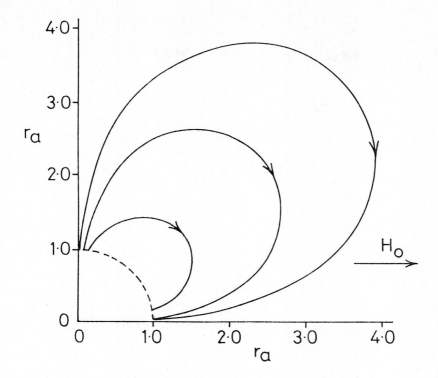

Fig.47. Projections, in the x-y plane of Fig.36, of particle
trajectories in the axial configuration.

symmetry near the cell boundary. Within the limiting radius, r_o,
which can be inscribed in each cell, the fluid velocity, $V(r)$,
rises from a zero value at the surface of the wire ($r=a$) to a
maximum value V_{ro} (where $dV(r)/dr=0$) at r . At all points outside
the limiting circle of each wire, the fluid is assumed to flow
at a uniform velocity, V_{ro}. This leads to a laminar flow pattern
thus allowing eqn.(48) for ideal flow in the z-direction to
be replaced with

$$\frac{dz_a}{dt} = \left(\frac{Pa}{2\eta}\right)\{ r_{oa}\log r_a + \tfrac{1}{2}(1-r_a^2)\} \tag{58}$$

where $r_{oa}=r_o/a$, where η is the fluid viscosity, and where P is
the pressure drop/unit length along the filter. Clearly, r_{oa} is
related to the filling factor, F, by $r_{oa}^2=\pi/2\sqrt{3}F$, for a hexagonal
close-packed array of wires. Integration of the velocity
profile of eqn.(58) over a unit cell, together with division by
the geometrical area of each cell, leads to a fluid flux/unit area,

G, such that

$$G = \frac{Pa^2}{8\eta} \{4-F-(2\log F+3)/F\} \tag{59}$$

Allowing for the volume of each cell occupied by a wire, this laminar flux density can be related to the ideal flow velocity, V_o, of eqn.(48) by

$$V_o = G(1-F) \tag{60}$$

The combination of eqn.(47) with eqn.(58) yields the time-independent equation of motion

$$\frac{dz}{d\theta} = -\frac{Pa^2}{2\eta V_m} \left[r_a^4 \frac{(r_{oa}^2 \log r_a + \frac{1}{2}(1-r_a^2))}{\sin 2\theta} \right] \tag{61}$$

which, along with eqn.(51) gives a complete description of the particle trajectory in the vicinity of wire, under conditions of laminer flow. The on-axis ($\theta=0$) particle capture trajectory, equivalent to eqn.(55) for the ideal flow case, is given by

$$z_a = \left(\frac{Pa^2}{4\eta V_m}\right) \left[r_{oa}^2 (\frac{r_a^4}{2} - Kr_a^2 \log r_a - \frac{r_a^4}{8} + \frac{Kr_a^2}{2} + \frac{r_a^4}{4} - \frac{r_a^6}{6} - \frac{Kr_a^2}{2} + \frac{Kr_a^4}{4} \right]_1^{r_{ai}}$$

Fig.48 shows a three dimensional plot of particle trajectories using eqns.(46,47) and (58) for the case where Vm/Vo=20 (ref.63). Here, the dashed envelope shown in the x,y plane of this diagram is the relevant portion of the isotelic curve. This curve is defined (ref.17) as the locus of starting positions (r_i, θ_i) for which capture occurs at the same distance L from the origin. In ideal flow corditions, capture length and capture time are synonymous and so the isotelic curve is seen here as the laminar flow equivalent of the isochronal curve. The isotelic curve in Fig.49 is truncated by the limiting normalized radius, r_{oa}, which is the typical situation in practical axial filters.

5.3 Single wire studies of particle build-up

A study of particle trajectories and of their arrival at the surfaces ($r_a=1$) of wires in the manner depicted in Figs.38 and 48 for the three principal capture configurations gives only a limited understanding of the behaviour of real filters. Clearly, what is required also is a knowledge of the stability of a particle on the surface of the fibre and of the manner in which

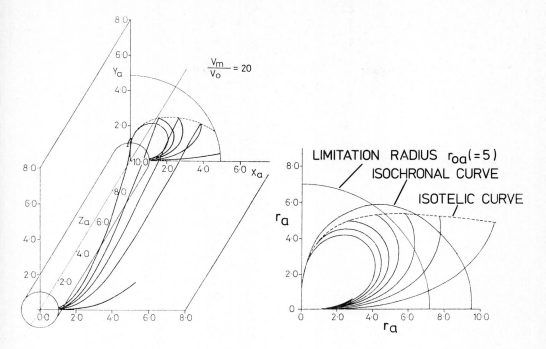

Fig.48. 3-dimensional plot of
trajectories of paramagnetic
particles in the axial configuration under conditions of
laminar flow. Here Vm/Vo=2.0
(ref.63)

Fig.49.Projection in the x-y plane
of Fig.48 of the isochronal curve
as well as part of the isotelic
curve (dotted line) for paramagnetic particles with Vm/Vo=10.
Both curves are truncated by the
normalized limitation radius,
$r_{oa}(=5)$. It should be noted that
when truncated by r_{oa}, the iso
chronal and the isotelic curves
give approximately the same
capture cross-section.(ref.63).

large numbers of particles accumulate on that surface.

From eqn.(39) it follows that, for conditions of ideal flow
and for a perfectly smooth fibre, the surface $(r_a=1)$ dynamics of
(say) paramagnetic particles in the L-configuration are given by

$$\frac{d\theta}{dt} = 2\left|\frac{V_c}{a}\right| \sin \theta - \left|\frac{V_m}{a}\right| \sin 2\theta \qquad (63)$$

In this configuration, paramagnetic particles will be stable only if $\frac{d\theta}{dt} < 0$. This means that the particle will be in stable equilibrium on the surface only if

$$\theta_{stable} < arc \cos \left(\frac{V_m}{V_o}\right)^{-1} \tag{64}$$

If this result is coupled with that of eqn.(32) describing the critical angle, θ, at which the radial force ceases to be attractive, then it can be seen that $2/(1-K)$ is the minimum value of $|V_m/V_o|$ for surface stability of paramagnetic particles in the L-configuration.

For $\frac{V_m}{V_o} >> 1$, the movement of particles over a smooth cylindrical surface, after arrival at that surface, is towards a position of stable equilibrium and is governed by

$$\theta(t) = arc \tan\{\exp \; (\pm \; \frac{V_m}{V_o} \; t)\} \tag{65}$$

where the $+$ sign is for diamagnetic particles, which migrate towards the $\theta = \pi/2$ position, in contrast to paramagnetic particles, which migrate towards the $\theta=0$ position.

For $|\frac{V_m}{V_o}| < \frac{2}{1-K}$, there arises, in the inviscid approximation, a sub-critical capture situation in which the particle arrives at, but is not held at, the front surface of the fibre. For example, in the case of a diamagnetic particle, in the T-configuration, with $|\frac{V_m}{V_o}| = 1$, eqn.(39) may be solved analytically. The solution

$$\tfrac{1}{2}(1-\sin\theta)^{-1}+\tfrac{1}{2}\log\{\tan(\tfrac{\pi}{4}+\tfrac{\theta}{2})\} = -2|\frac{V_m}{a}|t \; , \tag{66}$$

describes the surface movement of such a particle. The latter migrates over the surface in accordance with this equation until it reaches θ_c (eqn.(32)), at which point the radial force becomes positive and the particle is released from the surface.

In recent years there has been intense theoretical and experimental interest in the details of particle build-up on individual wires. The first major attempt to construct a theoretical model for this process was made by Luborsky and Drummond (ref.56). They studied paramagnetic particle build-up on ribbon-like fibres in the L-configuration assuming ideal flow conditions. In this calculation two alternative heuristic

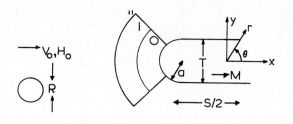

build-up geometries were assumed, the more successful of which is indicated schematically in Fig.50. Here, the assumption is made that for $V_m/V_o \gg 1$ the build-up profile fans out symmetrically in a 90° angular sector over the front face of the fibre such that, when n monolayers of identical particles (of

Fig.50. Build-up geometry predicted by Luborsky and Drummond for the paramagnetic particles on the L-configuration (ref.56).

radius R) are deposited, the fibre radius increases from the value a to a_n = a + 2Rn.

They then generalize eqns. (A18) and (A19) for ideal flow around a cylindrical fibre to the forms

$$V(r) = V_o(1-\frac{a_n}{r^2}) \cos \theta, \text{ and} \tag{68}$$

$$V(\theta) = -V_o(1+\frac{a_n}{r^2}) \sin\theta$$

For the simplified case of K = 0, the analytical solution for particle trajectories,given by eqn.(41) for a bare wire,takes the more general form

$$Y_{ai} = r_a \left[1-(a_n/ar_a)^2 \right] \sin\theta - (V_m/2V_o r_a^2)\sin 2\theta \tag{69}$$

leading to the result that the normalized capture radius r decreases with increasing particle deposition according to

$$r_{ac} = |V_m/2V_o| (a/a_n)^2$$

$$= |V_m/2V_o| (1+2nR/a)^{-2} \tag{70}$$

for the regime where

$$|V_m/V_o| \le 2^{\frac{1}{2}}(a_n/a)^2$$

This is obviously consistent with the theoretical results of Table 9. The limitation of this simple model lies in its inability to account for particle stability and, consequently, in its inability to describe a limiting build-up volume on

the wire surface for a given set of conditions. The same
authors, in a following paper (ref.64) improve greatly upon
this simple model by considering that the saturation value of
front-face build-up is determined by the locus of a zero net
azimuthal force acting on a particle in the build-up surface.
The two competing forces considered by Luborsky and Drummond
were the azimuthal component of the magnetic tractive force
$F_{M,\theta}^P$ (eqn.(29(B))) and the opposing Stokes drag force. The
simple drag force of eqn.(33)b was modified to include the
affects of the boundary layer (ref.65) created at the front
face of a cylinder by an impinging fluid. The thickness of
this layer, δ, averaged over the angles of interest

is given approximately as

$$\delta = \pi^{\frac{1}{2}} (\eta a_n / V_o \rho)^{\frac{1}{2}}$$

The quantity δ is larger in value than the typical particle
radius, R. This means that the azimuthal drag force of
eqn.(33(b)) is reduced by a factor R/δ, resulting in the modified
form

$$F_{D,\theta}' = 6R^2 (\pi \eta V_o^3 \rho / a)^{\frac{1}{2}} (1 + \frac{a_n^2}{r^2}) \sin\theta \qquad (72)$$

The condition $F_{M,\theta}^P = F_{D,\theta}'$ provides a limiting value of the
build-up radius at various angles, θ.

Under ideal flow conditions in the L-configuration, it is
possible to model the arrival of a particle at the downstream
edge $(\theta > \pi/2)$ of a cylindrical wire (refs.(64),(66)). In their
second paper, Luborsky and Drummond consider this particular
phenomenon. Here, they use, as a stability criterion, the
balance of the radial magnetic force (eqn. (29(a)) and the
radial component of the shear force, the latter being, once
again, modified by the presence of the boundary layer the value
of which is taken as (ref.65)

$$\delta \text{(rear surface)} = 5 (\eta S_n / V_o \rho)^{\frac{1}{2}}$$
$$S_n = S + 2nR$$

where S is the width of the ribbon-like fibre. This shear
force and the radial magnetic force are equated to provide

limiting angles of build-up on the downstream edge. An important
feature of this calculation is that rear surface build-up is
predicted as being of greater volume than that on the front face.

A more sophisticated iterative model of capture has been
introduced by Watson (refs. 66,67) in which a detailed
examination is made of the stability of particles arriving at
the wire surface. Watson introduces a friction-like azimuthal
force of the form $\lambda F_{M,r}$ acting on the paramagnetic particles

at the wire surface. Here, $F_{M,r}$ is of the form given in

eqn. (29(a)) and λ is a coefficient of friction which Watson
estimates, (on the basis of surface roughness produced by capture
particles of radius R), to be of order 0.3. This frictional
force enhances particle stability on the surface and assists
the azimuthal magnetic force, $F_{M,\theta}$ in opposing the hydrodynamic

drag. For the latter, Watson uses the same expression as Luborsky
and Drummond (ref.64). He then develops a statistical approach
to particle build-up in which the accumulation of particles
at any particular element of the surface is based on the
relative magnitudes of their surface velocities. In any element,
of the capture surface, the particles contributing to build-up
are of two types, namely
those arriving directly
at the surface element from
the fluid and also those
migrating from adjacent
surface elements. This
procedure is iterated
numerically six times by
Watson, each successive
iteration being carried
out with the flow conditions
modified by the presence
of freshly deposited
material. The build-up
profiles generated by this

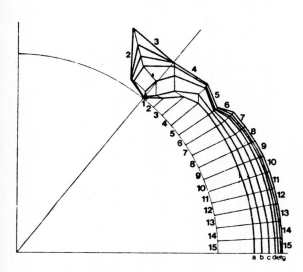

Fig.51. Profile of paramagnetic particle
build-up in the L-configuration computed
by Watson (refs.66,67)

model differ only in detail from those obtained from the simpler
model of Luborsky and Drummond. The particular point of variance
is that build-up material is present in the spatial region,
adjacent to the wire surface, in which $F^P_{M,r}$ becomes respulsive
(Fig.51). However, this model provides a considerable measure of
support for the predictions of the earlier simple model of
ref.56.

More recently, Nesset and Finch (ref.48) have refined the
calculations described above to include a more precise form of
the fluid drag force acting on a particle at the surface of
build-up. Here, for a fibre of cylindrical cross-section, they
consider the angular variation of the shear stress, $\tilde{\tau}_o$ acting
on stationary particles at various angles, θ, on the build-up
surface. Using the Blasius solution (ref.65) to the boundary
layer problem, they obtain

$$\tau_o = \rho_m (V_o^3 \nu/8r)^{\frac{1}{2}} B(\theta) \tag{73}$$

where $B(\theta) = 9.861\theta - 3.863\theta^3 - 0.0261\theta^7$
$$+ 0.0005\theta^9 + \dots \tag{74}$$

is the Blasius function, where ν is the kinematic viscosity
and where θ is defined from Fig.36. They assume the shear force
τ_o to act on an array of spherical particles, each of radius R,
sitting on the surface of the wire, after commencement of build-
up, such that the fluid drag on an individual particle is

$$F_{D,\theta}^{11} = f_R 4\pi R^2 \tau_o$$

where $f_R = \pi/8$ (ref.68) is a geometrical factor representing the
integrated effect of τ_o acting over the upper half of the
spherical particle. Nessit and Finch determine their outline
of saturation build-up by finding the locus of points at which
the total net azimuthal force acting on a particle is zero
as well as by finding the locus of points, close to the fibre
surface at which F_M^P is zero. They then calculate the maximum
value of the normalized build-up radius, (found to be
approximately constant over the angular sector $\theta = \pm\pi/4$) to be

$$r_a(\max) = \left[\frac{8^{3/2} \chi M_s H_o R \sin 2\theta}{3\pi B(\theta) a^{1/2} \rho_m V_o^{3/2} \nu^{1/2}} \right]^{2/5} \tag{75}$$

$$= N_L \left[\frac{8^{3/2} \sin 2\theta}{3\pi B(\theta)} \right]^{2/5} \tag{76}$$

where N_L is a dimensionless quantity, termed the loading number. Physically, N_L represents the ratio of magnetic forces to fluid shear forces at full load. Nessit and Finch show that N_L is related to V_m/V_o via the relation

$$N_L = \frac{9}{\sqrt{2}} \left(\frac{a}{R} \right) \left(\frac{1}{R_e} \right)^{1/2} \left| \frac{V_m}{V_o} \right| \tag{77}$$

where R_e is the Reynolds number of the wire. They then assume, for simplicity, that the angular half-sector of front face build up is exactly $\pi/2$ and proceed to show that the saturation volume loading of the fibre, γ_V expressed in terms of the volume of retained material per unit volume of fibre, is

$$\gamma_V = \frac{\beta}{4} \left[\left(\frac{N_L}{\xi} \right)^{4/5} - 1 \right] \tag{78}$$

where $\xi = 8^{3/2}/3\pi B(\theta=\pi/4)$ and

where β is the fractional packing density of the build-up material. The authors then find that, based on measurements with colloidal suspensions of CuO and MnO , this equation, with $B \simeq 0.7$, provides an acceptable description of saturation loading for paramagnetic material in the L-configuration. Given that, as eqn.(77) shows, N_L and V_m/V_o are directly proportional to one another it is of interest to note that, using a simple model similar to that of Luborsky and Drummond, Wong

(ref.20) has recently found an empirical relationship of
the form

$$\gamma_V = 0.137 \frac{V_m}{V_o}^{0.88}$$

for K = 0.8.

The phenomenon of rear surface build-up of paramagnetic
particles in the L-configuration at moderate $(6 < R_e < 30)$
Reynolds numbers has been treated theoretically by Watson
(ref.69). The flow field, at such numbers, around a
cylindrical wire is complicated (ref.70) and is characterised
by two vortices, symmetrically disposed, close to the rear
surface of the cylindrical wire (Fig.52). The mathematical
details of the calculation are beyond the scope of this
review but the following

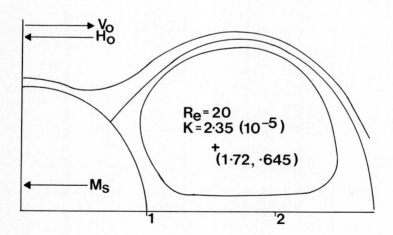

Fig.52. Stationary vortex downstream of a cylindrical wire
at moderate Reynold's number (after Watson ref.69)

points of summary may be made. He found that, with the above-
described flow conditions, the downstream capture cross-section
was always small and that capture was only possible for $V_m/V_o \ll 1$. He
also found that for higher values of V_m/V_o, the downstream capture
must be preceeded by front face capture, the particles being
swept from the build-up front to the rear. It was also found to
be necessary to include the inertial terms of eqns.(34) in order
to obtain capture trajectories in which the particles crossed
the streamlines of Fig.52. In this front-to-rear face migration,
the particle always remained close to the wire surface, since
V_m/V_o was very much less than unity. Watson was, however, unable to
calculate, a priori, a down-stream capture radius, nor the manner
in which the latter varied with particle build-up.

Fig.53 shows the results of a computation by Wong (ref.20)
using the inertialess particle trajectory equations (38) and (39),
together with eqn.(63) describing surface migration for a smooth
wire for paramagnetic particles (V_m/V_o = 2) in the L-configuration
Here, rear surface capture is achieved when (say) a particle
arriving from its limiting capture trajectory at the point
θ = 71.2^o migrates, in accordance with eqn.(63), across the
smooth surface to the angular position θ = 72.7^o.   In this
sector,the particle experiences a repulsive force at the
wire surface. The numerical solution of eqns.(38) and (39)
produces the trajectory indicated,in which the particle orbits
part of the downstream edge of the wire, arriving finally at
the angular position θ =148°.   Eqn.(63) confirms that all
particles, captured at this point, are able to achieve
stable equilibrium. Downstream capture, on the basis of this
simple model, is feasible.

Build-up in the axial configuration has been examined
theoretically and experimentally by Uchiyama and co-workers
(ref.71) in terms of the dynamics of the process. By a straightforward extension of the capture theory described above, for
the axial configuration, they show that the normalized on-axis
(θ=0) build-up accumulation radius, r_a, increases with time at
a rate given by

$$\frac{dr_a}{dt} = (t_M r_a^3)^{-1} \tag{79}$$

where $\quad t_M = \beta_p a / C_i V_m$

is a characteristic relaxation time and where C_i is the input
concentration of the particle suspension.

An extensive programme of experimental work on single-wire
particle build-up in all three principal capture configurations
has been carried out in the period 1975-81, much of it by
Friedlaender and co-workers (refs.57,72-77). In 1975, in the
first of their papers (ref.72), it was shown clearly, using
aqueous suspensions of MnO_2 and Fe_2O_3, that the build-up
of particles on initially clean fine ferromagnetic fibres

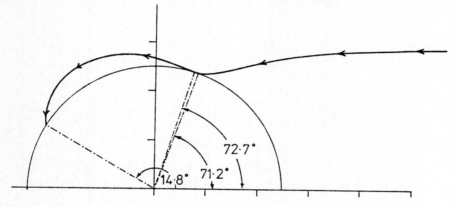

Fig.53. Surface migration of weakly-captured paramagnetic
        particles in the L-configuration (ref.20)

reaches an asymptotic limit characterized by a saturation radius
which is, typically, many particles in thickness. Fig.54 shows
a photographic time sequence obtained by Cowan et al (ref.72)
for the capture of MnO_2 particles on a 25μm diameter nickel wire
in the T-configuration. Unfortunately,the particle diameter
ranges used in these experiments (0.1 to 10μm for α-Fe_2O_3,
1.0 to 100μm for MnO_2) were too great to permit useful
quantitative analysis of build-up theory. Fig.55 shows a cross-
sectional build-up profile of monosized nickel-chromium alloy
particles on a coarse steel-wire in the T-configuration (ref.78).
The build-up of paramagnetic particles in this configuration and in
the axial configuration is characterized by two symmetrical outgrowths, in the vicinity of $\theta = \pm \frac{\pi}{2}$ in Fig.36, Similar outgrowths
occur in the case of diamagnetic particles captured in the L-
configuration (ref.14). By contrast, in the L-configuration, at

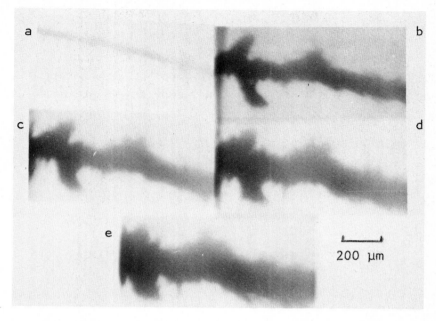

Fig.54. Capture of MnO_2 particles on a nickel wire in the T-configuration. The flow rate was 45mm s^{-1} and the entire time sequence a,b,c,d around 450s.(Courtesy of F J Friedlaender ref.72).

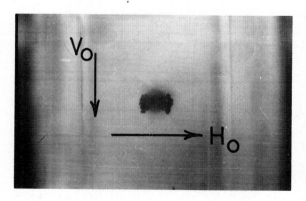

Fig.55. Cross-sectional view of cylindrical nickel chromium alloy particles captured on steel in the T configuration. (ref.78).

low Reynolds numbers (1) the build-up of paramagnetic particles
is confined largely to the front face of the wire in the
angular sector between $\pm\ \theta_c$ of Fig.36. This front-face build-up
strongly resembles that obtained for diamagnetic particles in the
T-configuration (Fig.2). Fig.56 shows experimental saturation
build-up profiles of $Mn_2P_2O_7$ on nickel wire obtained from photographic studies of Friedlaender et al (dotted curves) superimposed
upon the theoretical prediction (full curves) of Nesset and Finch
described above (ref.79).

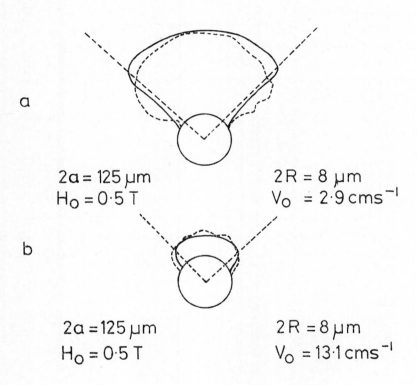

a

$$2a = 125\ \mu m \qquad 2R = 8\ \mu m$$
$$H_o = 0.5\ T \qquad V_o = 2.9\ cms^{-1}$$

b

$$2a = 125\ \mu m \qquad 2R = 8\ \mu m$$
$$H_o = 0.5\ T \qquad V_o = 13.1\ cms^{-1}$$

Fig.56. Measured (---) and calculated (——) build-up profiles
        of $Mn_2P_2O_7$ on nickel wire in the L-configuration for
        the values indicated using unpublished measurements by
        Friedlaender et al (after Nesset and Finch (ref.79))

In the axial configuration, Friedlaender et al (ref.73) have
shown, using various glycerol-water solutions as carrier fluids,
that the normalized radius of build-up, r_a, varies as time as

$$r_a^n = At + 1 \qquad\qquad (80)$$

where n is an exponent the value of which depends on the
Reynold's number and where A is a constant. The parameter n is
found, experimentally, to be equal to 4, in keeping with the above
mentioned build-up theory of Uchiyama and co-workers (ref.71).
The experimental constant A is found to be in agreement with that
theory only if the packing efficiency in the build-up profile lies
in the range 0.1 - 0.18. In addition, McNeese et al (ref.76) have
shown clearly that the normalized saturation radius, r_a, is (Fig.57)
proportional to $|V_m/V_o|^{1/3}$ in keeping with a simple balance model of
particle stability. The most recent experimental work by this group
(ref.77) is concerned principally with the monitoring of simultaneous upstream and downstream build-up in the L-configuration. As
in the original paper this work shows clearly (Fig.58) the presence
of saturation volumes of particle capture. However, a close study
of the data confirms that it is not possible to obtain a clear
relationship between this data and either of the above-mentioned
theoretical dimensionless groupings, $\dfrac{V_m}{V_o}$ (ref.47) or N_L

refs (48,79,80)   From this data it can also be seen quite clearly,
that the downstream build-up is always less than that on the front
face. In addition, they find that, at lower flow velocities, the
onset of downstream build-up is delayed for several minutes. At
higher flow velocities, more rapid downstream build-up is observed.
The mechanism mentioned above of front-to-rear face migration,
characterized by eqn.(63), is also observed.

5.4 Filter capture efficiency (recovery)
   The matrix of a real filter consists of a very large number
of individual fibres. It is possible, however, to define a capture
efficiency (or recovery) R for such a filter in terms of the value
of r_{ca} for an individual fibre and in terms of the fibre
distribution within the field volume. In the case of an ordered
longitudinal filter (with H_o parallel to V_o and γ=o for all fibres)
it is a simple matter to calculate this with a two dimensional
model (ref.63) analogous to that used to evaluate the mean free
path of molecular collisions in gases.

   Consider a particle travelling a large distance ℓ in an
infinite filter matrix. For any given set of conditions, a fixed
capture radius, r_c, can be assigned to any individual particle-fibre

252

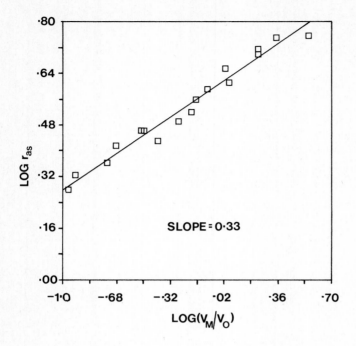

pair, such that, in its traversal, the individual particle sweeps out an effective area $2r_c$ within which any fibre would trap the particle. The volume filling factor of the fibres, F, that is the fractional volume of the filter occupied by the fibres, may be related to the number of fibres, n, within the swept area by $F = n\pi a^2 / 2r_c \ell$ where a is the fibre radius.

Fig.57. Linear relationship between saturation build-up radius and $(V_m/V_o)^{\frac{1}{3}}$ for paramagnetic particles in the axial configuration (after McNeese et al (ref.76))

In travelling a distance, ℓ, the particle would encounter n fibres capable of capturing it. The mean free path, $\bar{L}$, between captures is therefore $L = \ell/n$ and so the probability of escape by the particle in travelling a distance L, (where L is the length of a real filter), is $\exp(-L/\bar{L})$. Thus, the number of particles/unit volume, C_{out}, leaving a filter of this type is related to the number/unit volume, C_{in}, entering the filter by an expression of the type (ref.63).

$$C_{out} = C_{in} \exp(-L/\bar{L}) = C_{in} \exp(-\frac{2}{\pi} r_{ca} L_a F) \qquad (81)$$

where $L_a = L/a$,

and where $R = (C_{in} - C_{out})/C_{in} \qquad (82)$

Early theoretical expressions (refs.46,47) for recovery R, in random filters in the L-configuration assumed that the

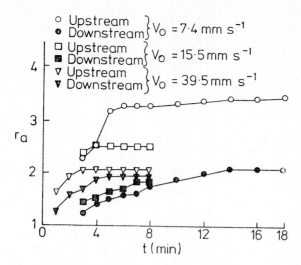

Fig.58. Upstream and downstream build-up of $Mn_2P_2O_7$ particles in a 50% glycerol-water slurry as a function of time. Magnetic field 0.5T applied in the L-configuration.(after McNeese et al ref.77)

distribution of wire orientations relative to H_o (and V_o) was equivalent to two thirds of the wires being ordered in the manner described above, (with the wires orthogonal to both field and flow), with the remaining one third being aligned parallel to the field and, as such, taking no part in the capture process. From this starting point, simple expressions for R, similar to eqns.(81) and (82) above, were derived.It has been realised for some considerable time that the expressions for R, calculated in this way, are an order of magnitude higher than the values obtained by experiment (refs. 56,81). T J Sheerer and the present authors have since considered (ref.82) the full implications of an isotropic wire distribution on the performance of a random filter operating in either the L- or T-configurations.

The general case where the angle between H_o and the wire axis, Z, is $(\frac{\pi}{2} - \alpha)$ has been considered (Fig.59) and all values of α_o in the range $0 < \alpha_o < \frac{\pi}{2}$ have been regarded as equally probable. The angle α_i of the internal field $\underline{H}_i$, (see Appendix 1), is similarly defined and depends on the relative magnitudes of $\underline{H}_o$ and $\underline{M}_s$ as well as upon α_o via a transcendental relation of the form

$$\alpha_i = \text{arc tan}(\sin \alpha_o/(\cos \alpha_o - K \cos \alpha_i)) \qquad (83)$$

where K is defined from eqn. (9). Fig.60 is a plot of $\cos\alpha_i$ and $\cos\alpha_o$ versus α_o for a value of K = 0.8. A further effect of arbitrary wire orientation is shown in Fig.61,where the motion

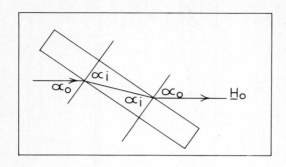

Fig.59. Direction of internal magnetic
field in a cylindrical ferromagnetic wire tilted at an
arbitrary angle to the
external field direction.

where K is defined from eqn. (9). Fig.60 is a plot of $\cos\alpha_i$ and $\cos\alpha_o$ versus α_o for a value of K = 0.8. A further effect of arbitrary wire orientation is shown in Fig.61, where the motion of the fluid relative to the wire axis is reduced from the value V_o in the simple theory above to a value $V_o \cos\alpha_d$ The effect of the non-orthogonal wire orientation has been shown, in this calculation, simply to replace V_m/V_o and K in eqns.(38) and (39) by more general coefficients

$$(V_m/V_o)^1 = (V_m/V_o)\cos\alpha_i \quad \text{and}$$
$$K' = K\cos\alpha_i/\cos\alpha_o$$

produced by the combined effects of magnetic depolarization and of reduced velocity between particle and wire. A further effect of this non-orthogonal wire orientation is to reduce the cross-section of wire presented to incident particles. If a wire captures particles initially offset from it a (normalized) distance r_{ac}, and is aligned at an angle $(\frac{\pi}{2}-\alpha_o)$ to the incident particle direction of motion (and hence to H_o) the (normalized) capture cross-section per unit length of wire, σ_{ca}, is

$$\sigma_{ca} = 2r_{ca}(\alpha_o)\cos\alpha_o$$

where $r_{ca}(\alpha_o)$ is a function of α_o obtainable from modified forms of eqns.(38) and (39). The mean value of σ_{ca} for an isotropic variation of α_o from 0 to $\frac{\pi}{2}$ over many wires is therefore

$$\sigma_{ca} = 2\int_o^{\pi/2} r_{ca}(\alpha_o)\cos\alpha_o\,d\alpha_o / \int_o^{\pi/2} d\alpha_o .$$

Since $r_{ca}(\alpha_o)$ is not analytic, numerical techniques are required to evaluate filter performance accurately. However, if weak capture is assumed, application of a linear dependence of r_{ca} upon $(V_m/V_o)\cos \alpha_i$ (see Table 9) yields the following expression for the mean value of r_{ca}.

$$\bar{r}_{ca} = r_{ca}(\alpha_o=0)\frac{2}{\pi} \int_0^{\pi 2} \cos\alpha_o \cos\alpha_i\, d\alpha_o$$

For K = 0.8 (say), this expression results in capture efficiency expression of the form

$$C_{out} = C_{in} \exp (-0.25 Fr_{ca} L_a) \tag{84}$$

It is of interest to note that, if magnetic depolarization effects are removed from this theory, (by making $\alpha_o = \alpha_j$) then the result

$$C_{out} = C_{in} \exp(- \frac{4}{\pi^2} r_{ca} L_a F) \tag{85}$$

is obtained, which is essentially that of the simple theories described above.

In the T-configuration, where H_o and V_o are orthogonal, the proper calculation of R is more complicated than in the case of the L-configuration because the obliquity of a wire with respect to H_o (or to V_o) is a three dimensional problem. It has been shown, nevertheless, that in this configuration, in the weak capture limit, with K = 0.8, the recovery is given approximately by the

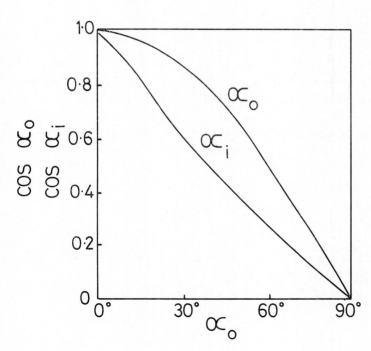

Fig.60. Variation of cos α_i with cos α_o for the case K = 0.8.

relation

$$C_{out} = C_{in} \exp{(-0.12r_{ca}L_aF)} \tag{86}$$

An important feature of this result is that, in contrast to the above-mentioned earlier theories (refs.46,47) but in keeping with experimental observation, it predicts a relatively weaker performance from T-configuration filters as compared with equivalent ones operated in the L-configuration.

The capture efficiency for an axial filter, comprising self-supporting parallel bundles of wires, can be obtained in the following way (ref.45). If n is the number of wires intersecting unit cross-sectional area of the filter, then the filling factor, F, is given by $F = n\pi a^2$. The probability that a particle penetrates the filter is given by $C_{out}/C_{in} = 1 - nA$ where Aa^2 is the cross-sectional capture area associated with an individual wire.

The capture area, Aa^2, is the area enclosed by the four symmetrical isochronal curves of eqn.(56). Integration of the righthand side of eqn.(56) with respect to θ between the limits 0 and $\frac{\pi}{2}$ (and multiplication by 4) gives an analytical expression for A such that

$$R = FK + \frac{4F}{\pi} (1 + \frac{(K-1)^2}{8L_{aM}} L_{aM})^{\frac{1}{2}} \tag{87}$$

$$R \simeq \frac{4F}{\pi} L_{aM}^{\frac{1}{2}} \tag{88}$$

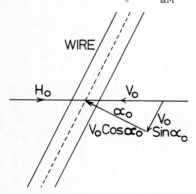

Fig.61.Reduced value $(V_o\cos\alpha_o)$ of the fluid velocity relative to the wire for arbitrary wire orientation relative to the field.

This simple theory is limited in two important respects. Firstly,for practical values of the magnetic length of the filter, L_{aM} and of thefilling factor, F, the capture efficiency, R, can assume values in excess of unity. In other words, this simple theory allows any initial value (r_{ai}, θ_i) for the

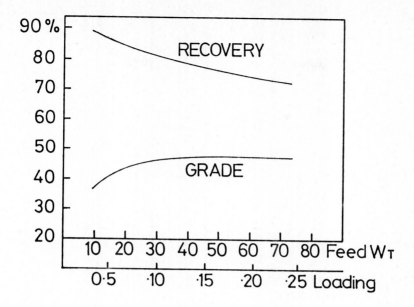

Fig.62. Variation of recovery and grade with matrix loading
for an HGMS system (after Mitchell et al (ref.8))

coordinates of the incoming particles, even those incompatible with
the inter-wire spacing. Secondly, this theory fails to take
advantage of the opportunity afforded by the comparatively high
symmetry of the axial geometry in terms of its description of the
fluid flow pattern around the wires. In this geometry it is a
relatively straightforward matter to introduce laminar fluid flow
conditions into the theory.

For a particle suspension passing through the axial filter
under conditions of laminar flow, the capture efficiency, R,
is measured by dividing the total volume of fluid swept clear
of particles in (say) unit time by the total volume passing
through the filter in that time.   Thus (ref.63)

$$R = \frac{2}{\pi} \cdot \frac{\displaystyle\int_0^{\theta_0} \int_1^{r_{oa}} r_a V_{ra} \, dr_a \, d\theta + \int_{\theta_0}^{\pi/2} \int_1^{r_{ai}} r_a V_{ra} dr_{ai} d\theta_i}{\displaystyle\int_0^{\pi/2} \int_1^{r_{oa}} r_a V_{ra} dr_a d\theta}$$

where θ is the angle at which $r_{ai} = r_o$ and where V_{ra}(eqn.(58)) is the fluid velocity at r_a.

The calculations of this section have assumed constant capture conditions for the filter. In fact, as the filter becomes increasingly loaded with particles, the recovery and the grade both change in value. This is seen to good effect in Fig.62 from the data of Mitchell et al (ref.8) for a Kolm-Marston separator system. The question of variation in performance with loading is dealt with fully in §5.7.

5.5 Physical capture

The only theoretical treatment of physical capture in random matrices is the phenomenological trial model of Luborsky and Drummond in 1975(ref.56). Here, they express recovery due exclusively to physical entrapment in the form

$$R \quad = \quad 1 \ - \ \exp \ (-f.FLr_{cp}/6a) \tag{89}$$

where r is a capture radius for physical entrapment and where the other symbols have been defined already. They then postulate that, based on experimental evidence, r should be proportional to $2(a+R)$ and proportional to $1/V_o$. This $1/V_o$ dependence is assumed from the increasing kinetic energy of the particle, which prevents sticking or which removes previously deposited particles through collision. Thus, they find that

$$r_{cp} \quad = \quad Q^1(a+R)/V_o a$$

where Q^1 is a dimensionless fitting factor.

Experimental measurements supporting this model were shown by Luborsky and Drummond (ref.64) in a following paper using $CuO-Al_2O_3$ slurries. Other published measurements supporting this general approach to physical entrapment have been made (Fig.63)by Clarkson and Kelland (ref.83) and by Dobby and Finch (ref.84).

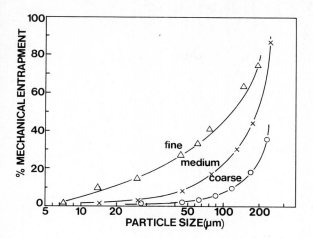

Fig.63. Mechanical entrapment in fine, medium and coarse filamentary matrices as a function of particle size. (After Clarkson and Kelland (ref.83)).

5.6 Flow resistance

Flow resistance to carrier liquids in filamentary matrices is very low. For a matrix comprising 50μm diameter fibres, the pressure drop across a 1cm matrix for water flowing at 1 cms^{-1} parallel to an edge is around 10^2 Pa. The authors and co-workers (ref.52) have described the flow resistance of filamentary matrices in terms of the classic Kozeny-Carman equation for porous media. This is done by treating the filter matrix as an array of cylindrical passages in parallel and by relating the average cylindrical radius to the porosity, ε, of the filter. It is assumed here that the passages are not linear and therefore the concept of an effective length L_e, is required, where L_e is greater than L, the physical length of the filter, such that the mean flow velocity is expressible as

$$\bar{V} = \frac{\Delta P}{\Delta \ell} \left[\frac{\varepsilon}{(1-\varepsilon)^2} \quad \frac{1}{\eta K^1 (S/V)_s^2} \right]$$

where S is the filter surface, V_s is the solid volume of the filter and K^1 is the Kozeny constant given by

$$K^1 = 2(L_e/L)^2$$

The specific flow resistance ξ of the filter, is defined as

$$\xi = . \frac{1}{V} \quad \frac{P}{L} = 4K^1(\eta/a^2) \ (1-\varepsilon)^2/\varepsilon^3 \tag{90}$$

The flow resistance of random matrices is found to be reliably described by eqn.(90). In Fig 64 the specific flow resistance of a variety of filters is shown as a function of filling factor, F. The experimental points are normalized for matrices constructed from 12μm, 25μm and 50μm diameter cylindrical fibre.

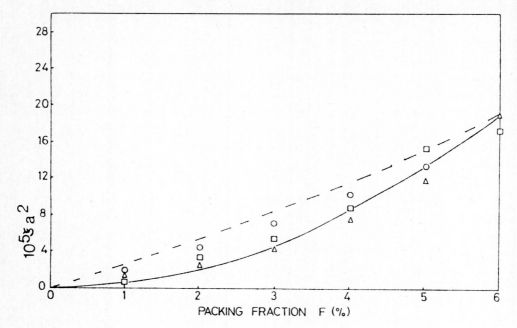

Fig.64. Normalized flow resistance in random filters as a
function of the filling factor for cylindrical wires
of diameter 50μm (○), 25μm(Δ) and 12μm(□). The solid
curve is the Kozeny-Carmen equation.

The continuous curve shows the Kozeny-Carmen equation fitted to
the data using a Kozeny constant $K = 11.3\pm0.5$. There is also
clear evidence of the validity of the $(1/a^2)$ dependence of flow
resistance upon fibre diameter.

As shown in §5.2, the authors and co-workers have evolved
for the axial geometry the only existing analytical expression
for flow resistance in terms of filter parameters.

5.7 Theory of breakthrough in magnetic filters
As has been already seen, the matrix of a high gradient filter
consists of a high porosity filamentary structure in which
particles undergoing filtration are deposited on to selected points
on the surfaces of the matrix material. The breakthrough in such
filters, brought about by the approach to saturation, has been
modelled theoretically in a successful way by assuming such
filters behave in the manner of fixed-bed adsorption.columns(ref.85).

In the following, a survey is made of the various analyses
(refs.81,85,86,87) carried out for a homogeneous absorption
filter, starting, first of all, with the simplest case of a
matrix of constant absorption coefficient α -   the so-called
strong-coupling limit (ref.86). This is followed by the more
general case in which α (x,t) is dependent upon the concentration of particles absorbed at a particular point in the filter
(x) at a particular time (t). The so-called (ref.86) 'weak-
coupling limit' where the length of the filter, L , is very
much less than $1/\alpha$ and where the absorption coefficient,
consequently, is a function $(\alpha(t))$, of time only, may be
seen to be one particular limit of the general case. Also in
this section, an examination is made of the absorption
characteristics of a filter capturing two (or more) distinct
species of particle (ref.88). Finally,axial filters, seen in §5.2
are examined in terms of the steady-state theory.

   Consider particles travelling through an absorbing filter. If
P(x) is the probability that a particle entering the filter at
x = O, is captured within a distance, x, then

$$dP(x) =   1 - P(x) \, dx$$

where α is the probability of capture/unit length of travel.
Thus

$$P(x) =   1 - \exp(-\alpha x)$$

and the probability that a particle passes through the filter to
the point x is now $1 - P(x)$ - that is $\exp(-\alpha x)$. The colloid
particle concentration C(x) can therefore be expressed as

$$C(x) = C_{in}\exp(-\alpha x) = C_{in} \exp(-x/L_o)$$

where L_o = $1/\alpha$ is the characteristic absorption length of the
filter (refs.81,86) where C_{in} is the particle concentration of the
input to the filter. The output concentration, C_{out}, of a filter
of length L is therefore seen to be

$$C_{out} = C_{in} \exp(-L/L_o) \qquad (91)$$

Clearly from eqn.(91) efficient filtration is not possible
unless $L \gg L_o$.

   During the process of filtration, particles are continuously
extracted from the fluid flowing through the filter on to the
capture surfaces. From first principles it is clear that the

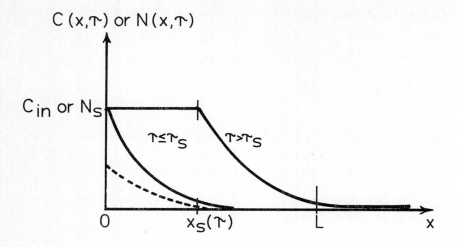

Fig.65. Schematic representation of carrier concentration
profiles $C(x,\tau)$ and build-up profile $N(x,\tau)$ for
values of normalized time greater or less (---)
than τ_s (after Collan et al (ref.81))

local concentration of captured particles $N(x)$ increases at a
rate

$$\frac{\partial N(x)}{\partial \tau} = V_o \alpha C(x,\tau) \qquad (92)$$

where $\tau = t - x/V_o$ is a renormalization time known as the
displacement time (which allows particle concentration and
build-up to be compared simultaneously at all points in the
filter) and where V_o is the entrance velocity of the fluid
to the separator. Also, from the equation of continuity (ref.87)
as applied to filters,

$$\frac{\partial N(x,\tau)}{\partial \tau} = -V_o \frac{\partial C(x,\tau)}{\partial x} \qquad (93)$$

Clearly, the number density of particles absorbed, $N(x,\tau)$,
increases at a rate proportional to the local concentration in
the fluid $C(x,\tau)$. If it is assumed, for the purpose of analysis,
that the filter is free of particles at the time $\tau = 0$, then

$$N(x,\tau) = V_o \alpha \tau \ C(x,\tau)$$

$$= V_o \alpha \tau \ C_{in} \exp(-x/L_o)$$

$$= N_o \ (0,\tau) \exp(-x/L_o).$$

For the strong-coupling limit, α ceases to be greater than zero throughout the filter after a time

$$\tau_s = (L_o/V_o)(N_s/C_{in}) \tag{94}$$

that is, when the absorption level at the front end of the filter $N_o(0,\tau)$ reaches a value equal to the saturation level N_s, of build-up on the matrix (Fig.65). At the front end of the filter α becomes zero; elsewhere it is constant. At time τ_s, both $C(x,\tau_s)$ and $N(x,\tau_s)$ decrease exponentially with position, x (see Fig.65)

i.e. $C(x_s\tau_s) = C_{in} \exp(-x/L_o)$ and   (95)

$$N(x,\tau_s) = N_s \exp(-x/L_o)$$

After the inlet of the filter becomes saturated, the length, x_s, of saturated matrix increases from zero with time and the active length of the matrix $(L - x_s)$ decreases. It is possible to write

$$x_s(\tau) = (C_{in}/N_s)V_o$$

The velocity $V_s(= dx_s/d\tau)$ of the saturation 'front' is $V_s = (C_{in}/N_s)V_o$.   Since x_s increases linearly with time, the colloid concentration profile $C(x,\tau)$ and the build-up profile remain exponential for filter positions x greater than x_s and move parallel to the profiles of the fresh filter, in a downstream direction.   Thus (Fig.65)

$$\begin{aligned} C(x,\tau) &= C_{in} \exp\left[(x_s(\tau) - x)/L_o\right] \text{ (a) and} \\ N(x,\tau) &= N_s \exp\left[(x_s(\tau) - x)/L_o\right] \text{ (b)} \end{aligned} \tag{96}$$

The time-dependence of eqns (96) is implicit in $x_s(\tau)$.   In the time period $0 < \tau < \tau_s$, the output concentration remains constant and of a value given by eqn.(91). Once the front end of the filter saturates (at $\tau = \tau_s$), then $C_{out}(\tau)$ starts to increase non-uniformly until, at a time $T = \tau_s + L/V_s$, that is

$$T = N_s(L + L_o)/V_oC_{in}$$

the saturation front reaches the output end of the filter, which is then fully saturated (Fig.65). For times $T > \tau_s$, the output concentration can be expressed as

$$C_{out}(L,\tau) = C_{in} \exp\left[(V_s\tau - L)/L_o\right] \tag{97}$$

or, alternatively, writing $\tau = n\tau_r$, where $\tau_r = \varepsilon_o L/V_o$ is the residence time of particles in the fluid, then

$$\log \ (C_{out}(L,\tau)/C_{in} \ = \ - \ (V_o/L_o H_o \varepsilon_o)(1 - \frac{\varepsilon_o nC_{in}}{N_s})H_o \tau_r$$

which agrees with empirical relationships in the literature (refs. 86, 89). If eqn. (92(a)) is differentiated with respect to time it is found (ref. 81) that

$$\frac{\partial C_{out}}{\partial \tau} \ (L,\tau) \ = \ (V_s/L_o)C_{out}(L,\tau) \tag{98}$$

This means that the characteristic absorption length $L_o (= 1/\alpha)$ can be obtained directly from eqn (98) via the time variation of C_{out}. It follows (ref. 81) that the level of build-up at the output end of the filter $N(L,)$ can be expressed as

$$N(L,\tau) \ = \ N_s \ \left[C_{out}(L,\tau)/C_{in} \right] \tag{99}$$

It also follows from eqn. (96) that for this simple model of build-up the absorption coefficient, α, (or the normalized capture radius r_{ca}) be replaced by the more general expression αG (or $r_{ca}G$) where (eqn. (97),

$$G \ = \ 1 - V_s\tau/L$$

$$= \ 1 - \frac{C_{in} \ V_o A}{N_s LA;}$$

(where A is the cross-section of the filter), i.e.

$$G(\tau) \ = \ (1 - N(x,\tau)/N_T),$$

where $N_T (=LAN_s)$ is the total number of particles held in the saturated filter. This simplified model is only approximately true in practice, even in extreme cases of filtration of highly magnetic particle suspension. In practice, G has been expressed (refs. 81, 87) as either

$$G(x,\tau) \ = \ 1 - (N(x,\tau)/N_s)^\gamma \quad (a) \tag{100}$$

or as $\quad G(x,\tau) \ = \ (1 - N(x,\tau)/N_s)^\gamma \quad (b)$

where, for the strong-coupling limit, γ has been found (refs. 81, 87) to have various values in the range $1 < \gamma < 2$ for a variety of filter types and of operating conditions.

Consider, now, the more realistic case where, after time τ_s, the absorption coefficient at the front end of the filter does not reduce completely to zero and where, at (say) the rear of the filter, the coefficient is less than that of the fresh filter at t = 0. Collan (ref.81) has dealt with this situation by assuming a staircase pattern for α in which its value depends upon the comparative level of build-up at a particular point in the filter. He has found that, even with a more involved absorption model of this type, eqns. (98) and (99) are still valid and that the breakthrough curve can still be used to interpolate the absorption pattern within the filter.

This less extreme type of absorption involving continuous change in $\alpha(x,\tau)$ has been investigated analytically by Uchiyama (ref.87). He assumes that one of the results of the strong-coupling limit described above is still valid for the general case, namely that the particle retention profile is fixed and moves downstream with increasing time. This he does by introducing a generalised coordinate, ξ such that

$$N(x,\tau) = N(\xi) \text{ and } C(x,\tau) = C(\xi) \tag{101}$$

$$\text{where } \xi = x - V_s\tau = (1 + V_s/V_o)x - V_s t \tag{102}$$

$$\text{or from eqn.(95)} = (1 + C_{in}/N_s)x - V_s t \tag{103}$$

Using equations (92) and (101), the continuity equation, eqn. (93), can be re-expressed as

$$\frac{d(N/N_s)}{d\xi} = \frac{d(C/C_{in})}{d\xi}$$

which, upon integration, gives the result

$$N(\xi)/N_s = C(\xi)/C_{in} \tag{104}$$

(This expression appears to be correct in that as the filter nears saturation and $N(\xi) \rightarrow N_s$, the fluid carrier concentration approaches that of the input (i.e. $C(\xi) \rightarrow C_{in}$).) Using eqns.(92), (93) and (104), Uchiyama obtains the expression

$$\frac{d(N/N_s)}{d\xi} = -\alpha\{N/N_s\} G(N/N_s) \tag{105}$$

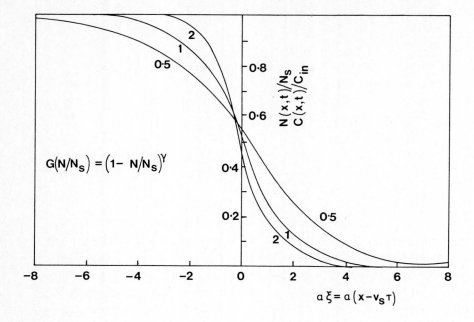

Fig.66. Calculated values of particle concentration and build-up
in random filters as a function of the parameter ξ
using the formulation of G expressed in eqn.(100(b))
(After Uchiyama and Kurinobu (ref.87).

where G is then given one of the forms of eqn.(100). This
equation can easily be solved analytically. If G is given the
form of eqn. (100) (a) then

$$N(\xi)/N_s = C(\xi)/C_s = \left(1 + A(\gamma)\exp(\gamma\alpha\xi)\right)^{-1/\gamma} \tag{106}$$

where $A(\gamma)$ is an integral constant dependent in value upon the
value of γ. For example, if $\gamma = 1$, $A(1) = 1$. Similarly,
$A(2) = 4$ and $A(3) = 27 \exp(\pi/2\sqrt{3})$.

Similarly, with the form of G given by eqn.(100)(b),it is als
possible to integrate eqn(105)analytically for certain values of
γ and to obtain a similar family of curves for this second alternative form. For $\gamma = 1$, the solution is identical to that of
eqn.(106) for $\gamma = 1$. For $\gamma = \frac{1}{2}$, it is found that

$$\frac{1 + \{1 - (N/N_s)\}^{\frac{1}{2}}}{1 - \{1 - (N/N_s)\}^{\frac{1}{2}}} = \exp(2 + \alpha\xi); \tag{107}$$

for $\gamma = 3/2$

$$\log \frac{\{1 + \{1 - N/N_s\}^{\frac{1}{2}}}{\{1 - \{1 - N/N_s\}^{\frac{1}{2}}} - \frac{2}{\{1 - (N/N_s)\}^{\frac{1}{2}}} = -2 + \alpha\xi \qquad (108)$$

and for $\gamma = 2$

$$\log (N_s/N - 1) - (1 - N/N_s)^{-1} = -4 + \alpha\xi \qquad (109)$$

The family of curves of N/N_s versus a ξ corresponding to the second formulation of G, is shown in Fig.66.

Clearly Fig.66 demonstrates the continuous variation in the absorption coefficient (or length) at (a) any point x in the filter at a fixed displacement time, τ, or (b) a particular fixed point in the filter as a function of time. A close examination of eqns.(106) to (109) and of Fig.66 shows that the Uchiyama analysis is only an approximate solution to the problem, even in the strong-coupling limit where $\alpha L \gg 1$. For example, in eqn.(106) for $\gamma = 1$ and for $\tau = 0$, it follows that $N(x,o)/N_s = 1/(1 + \exp(\alpha_o x))$ when, in fact, the right hand side should be identically 1. In fact, at the front end of the filter $(x=t=\xi=o)$, $N(o,o)/N_s=0.5$, with similar values (Fig.66) for non-unitary values of γ at $\xi = 0$. Clearly, the Uchiyama-Kurinobu theory islimited to usefulness in the strong-coupling limit, its principal limiation being the assumption that the profile of the breakthrough curve is fixed in shape throughout the life of the filter.

Consider, now, the case of the weak coupling limit, where, in the extreme limit, $G(=G(\tau))$ becomes independent of position within the filter and is a function of time only. In this situation, typified by the capture of weakly paramagnetic colloidal particles in fast process streams, $L \ll L_o$ and the filter bed becomes uniformly loaded. Here, it is possible to obtain highly simplified analytical forms for the time variation of G and therefore, of filter efficiency. Eqn .(92) can be simplified to the form

$$\frac{\partial L}{\partial \tau} \approx V_o \alpha G(\tau) C_{in} \qquad (110)$$

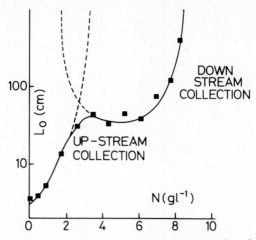

Fig.67.Variation in the absorption length of an
ordered filter in the L-configuration with build-
up.This variation is characterized by upstream
capture predominating in the early stages followed
by significant downstream capture at the later
stages.(After Collan et al.ref.88)

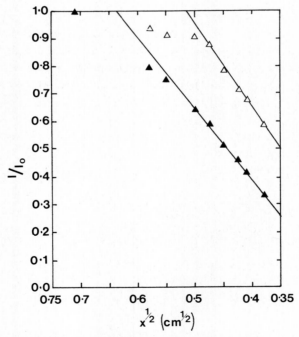

Fig.68. Characteristic breakthrough curve in an axial filter
obtained by monitoring the build-up of radioactive CuO tracer
particles (ref.90).

and, if G is expressed in terms of eqn. (100(a)), this is easily re-expressed by separation of variables, as

$$\frac{dG}{G^{\frac{2\gamma-1}{\gamma}}} = - \frac{d\tau}{\tau_0} \qquad (111)$$

where $\tau_0 = L_0 N_s / V_0 C_{in}$ is a relaxation time constant whose physical meaning may be regarded as the time taken by the saturation wavefront to move, (with velocity V_s), a distance equal to the characteristic length, L_0. For $\gamma = 1$, eqn. (111) leads to the result

$$G(\tau) = \exp(-\tau/\tau_0) \qquad (112)$$

a result obtained erroneously by Watson (ref.86). For $\gamma = 2$ $G(\tau) = 1 - \tanh^2 \tau/\tau_0$, using eqn. (100)(a) and $G(\tau) = (1 + \tau/\tau_0)^{-2}$ using eqn. (100)(b). Eqns. (111) and (112) may be seen to relate to the analysis of Uchiyama where, in the weak-coupling limit, where $x \simeq 0$, the quantity $\gamma \propto \xi$ in eqn. (106) becomes $-\gamma\tau/\tau_0$. Thus, for $\gamma = 1$, eqn. (106) becomes

$$N(\tau)/N_s = \left[1 + \exp(-\tau/\tau_0)\right]^{-1}$$

which, for $\tau > \tau_0$ approximates closely to the simple analysis of eqn. (112). Clearly, the weakness of the Uchiyama analysis becomes acute in the weak-coupling limit where negative values of displacement time are required to produce the full variation in $G(\tau)$.

Eqn. (110) is obviously inadequate for all but the weakest forms of capture. Clearly, an improvement can easily be made by replacing C_{in} by $C(\tau)$ where the latter is a suitable time-dependent average concentration in the filter. The Uchiyama value of $C(\tau)$ is $C_{in}(1 - G^{1/\gamma})$ which leads to the need for negative time. A simpler approach is to solve eqns. (92) and (93) giving, in the weak capture limit

$$\bar{C}(\tau) = C_{in}(1 - \alpha L G(\tau)/2) \qquad (113)$$

which leads to a solution of the form

$$\log G(\tau) + \{\log(1-R_0)^{\frac{1}{2}}\} \{(1-G(\tau))\} = -\tau/\tau_0 \qquad (114)$$

where $R_o (=\alpha L)$ is the initial capture efficiency of the filter. A somewhat stronger type of capture replaces C_{in} in eqn.(113) with $(C_{in}/\alpha LG(\tau))$ $|1 - \exp(-\alpha LG)|$, resulting in

$$G(\tau)+(1/\alpha L) \log \{\frac{1 - \exp(-\alpha LG(\tau))}{1-\exp(-\alpha L)}\} = 1 - \tau/\alpha L_o \qquad (115)$$

Expressions for $G(\tau)$ for $\gamma > 1$ can similarly be generated.

Recent work by Collan et al (ref.88) has described investigations of two further complicating factors in filtration theory. The first of these is concerned with two competing capture mechanisms within a single filter, such as front and rear side capture in the L-configuration. This is treated by assuming that the filter is composed of two interpenetrating sub-filters, each with its own absorption coefficients α_n (n=1,2). Here

$$N(x,\tau) = N_1(x,\tau)+N_2(x,\tau).$$

In the strong coupling limit the two absorption coefficients are shown to be able to be combined into a single two step absorption coefficient $\alpha(N,\tau)$. Fig.67 shows experimental results of Collan et al. for a parallel wire filter in the L-configuration in which it is seen that, early in the life cycle of the filter, normal front face capture dominates, and, as this type of capture reduces rapidly, rear face capture starts to increase, producing a more complex form of $\alpha(x,\tau)$ than any described above.

One further complicating factor considered by the same authors is that of polydisperse suspensions where the input concentration C_{in} is replaced by $C_{in,i}(i=1,...n)$. Similarly, α, N are replaced by α_i, N_i. Their analysis shows that, in the strong coupling limit $\alpha_i(\tau) >> 1$, at a distant point from the filter $(\alpha_i L >> 1)$ can be described, superficially, in the same way as a filter absorbing only particles of one species.

The axial filter has been shown earlier to have special absorption characteristics, which may be expressed as (eqn.(87)).

$$C(x,\tau) = C_{in} |1 - (x/L_o)^{\frac{1}{2}}| \qquad (116)$$

Filters of this sort exhibit a behaviour pattern in which the

front end of the filter quickly becomes saturated ($N_{in}N_s$)   and
a saturation front advances along the filter. The output
(x=L) concentration is then given by

$$C(L,\tau) = C_{in} \{1 - |(L - x_s)/L_o|^{\frac{1}{2}}\} \qquad (117)$$

where x_s represents the position of the saturation point in the
filter. The capture efficiency of the filter is expressed as

$$R = (C_{in} - C(L,\tau))/C_{in}$$

$$\doteqdot (L/L_o)^{\frac{1}{2}}(1-x_s/L)^{\frac{1}{2}} \qquad (118)$$

Now $(L/L_o)^{\frac{1}{2}}$ represents the capture efficiency, R_o L.
Therefore, as build-up proceeds along the filter length, the
capture efficiency varies as $(1-N/N_\tau)^{\frac{1}{2}}$ where N_τ is the total
number of particles which can be held in the filter. Fig.68
shows, using radioactive tracer techniques (ref.90) the advance
of this saturation point in an axial filter comprising 12μm
diameter fibres at 8% packing fraction (ref.90). Clearly, Fig.68
bears a strong resemblance to the schematic breakthrough curve
of Fig.65 for a random filter in the strong-coupling limit. Here,
however, the logarithmic relation between N/N_s and x is
replaced by a unique ½power dependence which strongly characterises
this type of filter. In a practical cyclic magnetic filter system,
of the type shown in Fig.32, the capture efficiency, R, will,
ultimately fall to a value which is unacceptable. At this point,
the flow is halted, the magnetic field switched off and the
filter cleaned by wash water. The magnetic field is then switched
on and the flow re-started,   For such a separator, the
processing rate, p, can be expressed as (ref.69)

$$p = V_o A \, N_o/(N_o + 1 + D/\tau_r)$$

where N_o is the number of canisters of fluid prcessed for which
an acceptable value of R can be obtained, A is the canister
(i.e. matrix) cross-sectional area and D is the time taken to
switch the field on and off and to flush the canister.

5.8 OPEN GRADIENT MAGNETIC SEPARATION

At the time of writing this review, there is a growing awareness
of the increasing importance of open gradient separation devices.

In an open gradient (OG) separation device (which contains no matrix) magnetic field gradients produced by current-carrying magnet systems deflect particle beams in directions which depend largely on the mass suceptibility of the materials involved. The advantage here is, of course, that the device operates continuously. In a sense this is not a new idea: the Frantz isodynamic separator described in §4.5, particularly when working in the free-fall mode, may be regarded as the arche-typal open gradient separator. A certain amount of experiment-ation has been done on OG separation using superconducting magnets, particularly quadrupole magnets such as that used by Kolm in 1968 (refs. 91,92) for the beneficiation of Mo ores. As seen in Fig.69, this system has, along its axis, a pair of concentric pipes connected by perforations along the surfaces of the inner pipe. Magnetic particles carried in a slurry in the inner pipe slowly diffuse radially outwards through these perforations under the influence of the quadrupole field gradients. A somewhat similar device also involving a superconducting quadrupole system, has also been used by Cohen & Good (ref.93) for the separation of denser mineral slurries. Here, an annular pipe is wound in a spiral fashion around the outside of the cylindrical magnet system. By generating orbital slurry flow in the pipe, all parts of the slurry are made to pass through the regions of strongest magnetic field gradient. By this means, high susceptibility particles are held at the pipe wall adjacent to the magnet. A layer of slowly moving magnetic particles of a few millimetres thickness is eventually dis-charged (separately) from the exit of this annular pipe. More recently, it has been reported (ref.94), that a new OG super-conducting device sponsored by the D.O.E. has been designed by E C Hise and co-workers

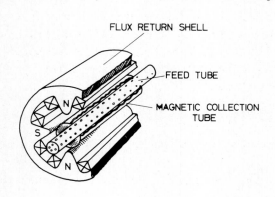

QUADRUPOLE MAGNETIC SEPARATOR

Fig.69. Superconducting quadrupole open gradient separator (ref.91)

at Oak Ridge National Laboratory. This comprises a large
helmholtz pair with the coils driven in opposition. Each field
coil has a 90 mm room temperature bore, 0.3 m length, and has
a field rating of 8.0T. It is intended that dry coal powder
be fed, under gravity, through the vertical axis of the magnet
pair. This system thus resembles strongly the Frantz isodynamic
separator in its free-fall mode of operation.

6. APPLICATIONS OF HGMS

A detailed survey of the developing applications of HGMS is
beyond the scope of this review. However, in the following, a
summary is given of some of the more successful areas of
innovation. As indicated in Table 2, the developing applications
of HGMS devices using filamentary matrices are numerous. Within
this large diversification, there have appeared in the last
decade three large-scale applications of particular note. These
are the beneficiation of kaolin clay, the removal of inorganic
sulphur from powdered coals and the treatment of industrial waste
water.

6.1 Beneficiation of kaolin clay

Currently, the most successful widespread industrial application
of filamentary matrix systems is the purification of kaolin clay,
a mineral of vital importance in the quality paper manufacturing
industry. This application has been discussed in some of the
afore-mentioned reviews (refs.8,9) and in a variety of commercial
and technical (refs.89,95) publications. This purification enhances
the brightness of the clay, a naturally white aluminosilicate,
thereby (a) producing higher quality clay than previously
obtainable by traditional clay 'whitening' methods, (b) making
available new clay reserves previously regarded of unsatisfactory
quality, (c) making existing waste deposits available for treatment
and (d) reducing the demand for industrial bleaches in the clay
industry. The brightness of kaolin is enhanced by the removal of
micron-sized mineral impurities whose identities depend to a great
extent on the geographical location of the clay deposits. Such
impurities include mica, iron pyrite, quartz, tourmaline, and
titanium dioxide. Most of these impurities are paramagnetic and
can be separated from a dense wet slurry ($\sim 20\%$ solids) of
kaolin in a Kolm-Marston type of separator. Installations

of the type shown in Fig.33 have been operating in Georgia, U.S.A since 1971.   All large-scale commercial installations are cyclic devices operating in the manner of Fig. 32.

 Oberteuffer and Wechsler (ref.96) have cited typical operating parameters for such a cyclic system operating on clays. These include a 12.5 min matrix loading time and a slurry velocity of 17 mms^{-1}, for clays containing 8gl^{-1} of mags in the feed. Flush times are typically of the order of 60s.

Industrial clay processing flowsheets follow the general pattern shown schematically in Fig.70 , where the inclusion of HGMS is shown in three alternative locations. In the upgrading of crude clay feeds, the slurries may be treated by HGMS prior to bleaching (A). Alternatively, it may wholly replace bleaching (B). In the production of high-value, high brightness clays, the HGMS system may be placed in the final stages of processing (C).

It is not possible to determine a single process cost/tonne for clay. Costs are dependent on the particular clay under consideration and upon the degree of upgrading required.

6.2 Water Treatment

The advent of the Kolm-Marston HGMS filter of Fig.32 has caused a considerable interest to develop in the large-scale treatment of both municipal and industrial waters. Processes which have evolved include direct filtration of magnetic particulates from water as well as indirect filtration, with the use of a magnetic scavenging agent, of water contaminated with non-magnetic particles.

The direct treatment of steel mill waste and process waters is an obvious area of application of HGMS. Up to around 150m^3 of water are required to process 1 tonne of steel and not all of this water is fit for recirculation within the steel making process since the waters may be heavily contaminated. One major advantage of the Kolm-Marston filter in this environment is that its high rate of throughput for waste waters is coupled with a relatively modest space requirement.

 Cyclic, automatic backflushing filters, as shown schematically

in Fig.32 , and having commercial specifications as shown
in Table 8, are now used widely. They operate in alternate feed
and flush modes. A typical duty cycle is one or more hours of
feed time to fractions of a minute for cleaning. Backflush
volumes are modest and are characterized by rapid settling (ref.97).
Quasi-continuous operation is possible using the feed surge tank
principle.

In the modern steel industry there are a variety of prime
sources of water contamination. These include the gas scrubber
waters from blast furnaces (BF), basic oxygen (BO) furnaces
and from steel de-gassing operations. They also include streams
from coking, casting and plating processes (which are generally
flocculated with streams containing magnetic particulates
from other steel making operations). Cooling waters from
continuous casting operations are contaminated, as are
the waters from hot forming and cold rolling operations.

As early as 1975 Oberteuffer et al (ref.97) made a
study of BF particles $(Fe, FeO, Fe_2O_3, Fe_3O_4)$which had a median
diameter of around 10µm and which were found to be less
magnetic than particles produced in the BO furnaces.and '
accounted for by higher measured levels of silica conta -
mination in the former. They also found scale pit waters heavily
contaminated with iron and iron oxide flakes, a significant
proportion of which were colloidal. Water-oil emulsions
used as a lubricant in cold rolling mills are known to
contain, with the oil, microscopic iron and iron oxide
particles which adversely affect the surface of finished
steel plate. Conventional technology for cleaning up these
waters involves, in all cases, large volume settling
tanks, clarifiers and deep base sand filters. Such traditional
methods are not capable of producing satisfactory effluent
and simply manage to make possible water recirculation within
the steel plant.

In 1975 a consortium of six U.S. Steel companies was formed
(ref. 97) to support a pilot and demonstration test programme
to apply Kolm-Marston filtration techniques (HGMF) to the steel
industry. In this programme, tests were performed on the

overflow waters from a scale pit treating water from a hot
rolling mill. The results of this test programme, using a 500gpm
SMI system were :

	untreated	treated	reduction
Suspended solids	$63 mgl^{-1}$	$4.4 mgl^{-1}$	93%
Oil	$9.8 mgl^{-1}$	$5.9 mgl^{-1}$	40%

These data were obtained at a filter surface loading rate
of $200 gpm/ft^2$ in a magnetic field of 0.5T with a cycle
period of 90 min. A variety of on-site tests have been
made by Sala Magnetics Inc. in the U.S., Sweden and Japan in
the period between 1976 and 1979.

Following the test programme, described above, the first
full-scale HGMF system was installed in a steel mill complex
at Kawasaki Steel Co., Chilsa, Japan. This was a 2.0m bore
SALA 214-15-5 HGMF system. It is important to note that this was
the first industrial scale Kolm-Marston filter system to be
introduced apart from those employed in the beneficiation of
kaolin clay. This was, in fact, used to filter recirculating
scrubber water from a steel de-gassing process. Details of
performance, found to be satisfactory at a field rating of
0.3T, have been reported by Takino et al (ref.98). With a
system throughput of 100 gpm/ft and with a cycle time of 1h,
they obtained the results :

	Untreated	treated	reduction
Suspended solids	$84 mgl^{-1}$	$10 mgl^{-1}$	88%

The matrix was cleaned by a backflush in zero field at
a rate of $400 gpm/ft^2$.

In 1978 Yano and Eguchi (ref.99) published HGMF results
on similar types of steel mill waste water. They found
in experiments carried out at Daido Steel that waste water
from the scrubbers of an oxygen converter contained contaminants
60% of which were magnetic. This scrubber water also contained

Ca ions at 120ppm which caused scale growth problems on the matrix.
They obtained upto 99% efficiency in the removal of suspended
solids at an input density of 200 ppm, at a flow rate of
200mh^{-1} and at a field of 0.3T. They also filtered water
from the hot rolling mill, which was found to contain
contaminants 30% of which were magnetic. These particles
were found to be associated with oil droplets at 5 to 15ppm and
were removed from the stream on to the matrix. The oil was
cleaned from the matrix periodically using high temperature
steam. HGMF at 600mh^{-1}, at 204ppm input density and at a field
of 0.3T was found to be over 90% efficient. They also treated
waters from a vacuum degassing installation containing suspended
particles only 10% of which were magnetic. Here, it was found
desirable to introduce a polymeric flocculant. Under similar
operating conditions they found that the addition of only 0.1 ppm
of polymeric flocculant increased filtration efficiency to nearly
80%.

Both conventional and nuclear power installations are
associated with steam generating systems using recirculated water
of high purity. In addition, in the United States alone, there
are over one hundred PWR and BWR power installations which have
their own high purity light water reactor cooling cycles. Low
levels of impurities found in such cycles include microscopic
traces of various oxides of iron as well as certain dissolved
solids. Such impurities contribute, among other things, to
radioactivity from impurity build-up in reactor circuits,
sludge build-up in steam generators, turbine blade build-up
and magnetite build-up in heat exchangers. Water cycles
of this type have been treated reasonably successfully (refs.38)
for several years using commercial electro-magnetic filters
of the type shown in Fig.24.   However, HGMF plant has shown a
greater ability to remove not only all iron species to ppb levels
but also to remove radioactive transuranic compounds and also
troublesome traces of Cu and CuO.   Such reductions are even
accompanied by level reductions in the undissolved salts
of Ca and Mg.

Schematic flow diagrams of the inter-related reactor and
steam generator cycles of (say) a PWR power installation are shown
in Fig.71. . Here, labels 1 and 2 indicate the passage of high

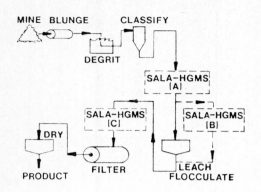

MINE BLUNGE   CLASSIFY

DEGRIT

SALA-HGMS
[A]

SALA-HGMS   SALA-HGMS
[C]   [B]

DRY

PRODUCT   FILTER   LEACH
                     FLOCCULATE

temperature pressurized
water from the reactor core
to the primary heat
exchangers in the (upper)
steam generation cycle.
Labels A through L indicate
feasible points of installation of HGMF plant. The
ambient conditions at these
various points vary
considerably, ranging from
75 psi and 320 F at E(say)
to 2500 psi and 650 F at A.

Fig.70. Alternative inclusion points
(A,B,C) for HGMS systems in
the treatment of kaolin clay.
(Courtesy of Sala Magnetics
Inc.)

In this area of appl-
ication the cyclic automatic
backflushing filters of the
type shown in Table 8
generally operate with typical duty cycle feed times of days or
even weeks. Flushing procedures as in other applications are
typically of the order of one minute.

Since 1976, various pilot tests have been performed on several
thermal power systems in the United States (refs.96),and full-scale
industrial installations began in 1978. Since 1978, three major
Sala Magnetics Inc. HGMF systems have been installed at paper mills
in the Southern U.S.A. for the treatment of steam condensates.
Specimen results for one such installation (ref.69)   are

Filter surface loading	200 gpm/ft^2	400 gpm/ft^2
Background field	0.5T	0.5T
Total Fe removal	89%	73%
Cu removal	57%	50%

The potential for HGMF in PWR systems has been fully discussed in a
published report submitted to the EPRI (ref.100). It has been shown
feasible for HGMF to be installed in the primary pressurized water
system for improved containment of radioactive species (ref.101).

It has been realized for some time (ref.102) that HGMF may be
used to remove colloidal diamagnetic particles from waters by

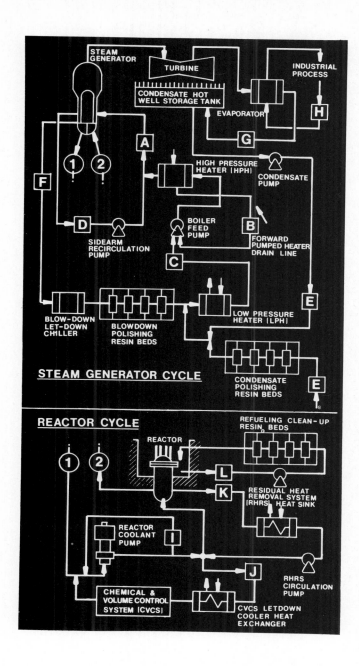

Fig.71. Schematic representation of reactor and steam
generator cycles, with labels 1 & 2 indicating passage of high
pressure, high temperature water. Labels A through L indicate
feasible points of installation of HGMF equipment.(Courtesy
of Sala Magnetics Inc.)

causing them to associate with a strongly magnetic seed
material (typically colloidal Fe_3O_4). This association is
brought about by the addition of an appropriate flocculant (at a
level of several ppm) such as alum $(Al_2(SO_4)_3)$ or one of several
commercial polymeric flocculants.

In 1973, De Latour (ref. 102) demonstrated the feasibility of
using HGMF to purify waters at high speed using a seeding
technique. His measurements were concerned with suspended
inorganic particle removal and with dissolved phospherous removal
both from sewage and from contaminated river water. It has also
been found possible to remove coliform bacteria (ref.102) and
even viruses and single cell protein (ref.103). A process design
for magnetite-seeded HGMF was developed by Sala Magnetics Inc.
in 1978. A 200 gpm demonstration system of this type has been
built by them and used at various sewage treatment plants
in Sweden to evaluate the effectiveness of HGMF for phosphorous
removal (ref.96).

Petrakis and Ahner (ref.104) have also investigated the
use of HGMF magnetic seeding applied to oil-water mixtures
from chemical sewers containing 175-1400 ppm oil and 29-320 ppm
of suspended particles. These investigations indicated clearly
that finely divided magnetite was much more effective than coarse
grain in obtaining high levels of recovery. They also showed
clearly that, as the flow rate through the matrix was increased
beyond a critical value, the magnetite seed particles were
captured but that drag forces sheared some of the contaminant
flow from the seed particles. In these experiments, alum-
generated flocs showed superior sheer resistance to those
generated by polymers. One further successful finding of this
work was that seeding techniques, with magnetite added at 300 ppm,
can be used to reduce levels (by 96%) of polychlorinated biphenyls
(PCB) in waste waters.

6.3 Desulphurization of Industrial Powdered Coals

Sulphur is present in coal in three main forms. The first is
organic sulphur, in which the sulphur is dispersed atomically
within the crystallites of coal. The second form is pyritic
sulphur, in which the sulphur is bound chemically to iron in the

form of pyrite (FeS_2) or marcasite, pyrrhotite (FeS_x, where   1.4)
or other sulphide forms. The third form is sulphate sulphur,
in which sulphur oxides are bound chemically to iron, calcium or
magnesium. Organic sulphur obviously cannot be extracted by
magnetic separation or by any other physical process, because
of the atomic nature of its association with the crystallites of
organic coal. On the other hand, inorganic forms of sulphur, once
liberated by crushing and grinding, can be processed by various
physical means including magnetic separation.

In 1958, a report was published (ref. 105) of the first
major attempt to clean finely ground coal by magnetic separation.
In this project, the coal was pretreated with superheated steam
and air in an attempt to stimulate thermo-chemical transformation
of the surface layers of the inorganic sulphur from the weakly
paramagnetic pyrite form into various more strongly magnetic
forms of magnetite, haematite and pyrrhotite. The results of this
early study were supported by similar successful experiments on
dry powdered coals by Kester (ref.41) using the Frantz isodynamic
separator (Fig. 26) and by Harris (ref.106) using the induced
roll separator. In 1968, Ergun and Bean demonstrated the
possibility of using r.f. fields to heat selectively (and thereby
transform) pyrite in powdered coals by making use of the
fact that the electrical conductivity of pyrite is much higher
than that of pure coal.

In recent years, such studies have expanded greatly with
the advent of the Koln-Marston HGMS system. The initial work
with such systems was carried out on bench-scale apparatus using
water-based slurries of powdered coal. Initial results were
published in 1973 by Trinidade and Kolm (ref.107). Using a
magnetic field of 2.0T, stainless steel matrices with average
fibre diameter of 100μm, and filling factors in the range 1 to 13%,
they obtained up to 80% recovery of inorganic sulphur from the
feed. The feed comprised particle suspensions with 25 to
30% solids by weight and particle sizes between 40 and 350 mesh.
At this level of recovery the slurry speed in the matrix was $4 mms^{-1}$
Since then, a variety of large-scale investigations have been
undertaken into the use of HGMS in coal processing, many of these
being funded by the United States Department of Energy (DOE).

It should be meade clear at this point that the successful
treatment of industrial coals on any practical scale is a
problem of considerable magnitude. First, the variable
nature of industrial coal means that it is not possible to
specify a single set of optimum particle size fractions and
of operating magnetic field values. The feed supplied to
the separator contains the organic ash-free and non-
pyritic coal. It also contains magnetically polarizable
free particles of ash and of inorganic sulphur which can
be trapped relatively easily at capture sites on the matrix.
Furthermore, it contains middlings comprising particles
containing both organic coal and unliberated paramagnetic
inorganic materials. If too high a field is used in the
separator system, some of these middlings are trapped
on the matrix, thus causing an important fraction of the
inorganic coal to be included in the mags, with a consequent
reduction in the thermal (Btu) value of the system product.
This is, of course, a clear example of the regression between
recovery and grade described in general terms in §2.4.
Secondly, the major difficulty with the wet processing of coal
in aqueous slurries on any industrial scale lies in the economic
cost of the dewatering of, and the subsequent drying of, finely
crushed coal. Consequently, it is fair to say that interest in the
magnetic treatment of aqueous slurries of coal has diminished in
relation to other schemes outlined below.

As an alternative to water-based slurries, wet HGMS has
been carried out using methanol (ref.108) and fuel oil
(ref.109) as carrier liquids. Both of these liquids have lower
interfacial tension with the coal particles than water and, as
such, can be more easily dispersed. In both cases, also, the
carrier liquid can be re-cycled after sedimentation of the
product following separation. Furthermore, drying is
unnecessary since residual liquid remaining with the coal
can be burned. The work of Maxwell and Kelland (ref.108)
with methanol and the DOE-funded fuel oil project at the
Naval Ordnance Station (Indian Head) (ref.109) showed
inorganic sulphur and ash removal comparable with the results
for water-based slurries.

Considerable attention has also been given to another
wet HGMS process, the magnetic treatment of solvent
refined coal (SRC). The SRC process involves pulverization
of industrial coals followed by their solution in a
volatile coal-derived organic solvent. Following
desulphurization by HGMS, the solvent is removed from the
product by flash evaporation. Initial investigations
with HGMS in SRC process streams were carried out by
Hydrocarbon Research Inc. (HRI) (ref.110), followed by
work of a similar nature by Liu and Lin (ref.111) and by
Maxwell (ref.112). The HRI work (ref.110) was carried
out with liquified coal at 350°F using liquid flow rates
of $0.14ms^{-1}$ in a stainless steel matrix in a field of
2.0T. These investigations succeeded in removing 90% of
the inorganic sulphur and 25 to 35% of the ash. Similar
results were obtained by Liu and Lin using liquefied coal
at 300°F. The MIT work (ref.108) on SRC streams was
carried out over a wider range of temperatures (100 to
575°F) and fields (2.0 to 8.0T) than preceding investigations.
It was found here that liquid stream temperatures in the
region of 430°F were optimum in terms of the overall
efficiency of the SRC process. At this temperature, using
magnetic fields of 2.0T, flow velocities of $26mms^{-1}$ and
a matrix 7 inches in length, it was found that essentially
all of the inorganic sulphur and up to 40% of the ash was
removed. This work showed that the optimum temperature is
closely linked to structural phase changes in pyrrhotite
(Fe_7S_8)where hexagonal forms change to monoclinic in the region of
220°C (ref.113). The M.I.T. project also demonstrated clearly
the feasibility of backwashing almost all of the inorganic
sulphur and ash from the matrix in zero field.

There are, of course, obvious advantages to the dry
magnetic separation of coals and a considerable amount
of work in this area has been funded by the D.O.E. The
first major project along these lines was carried out
by GE (in association with M.I.T. and the coal industry)(ref.114).
These results were disappointing though an important finding,
confirmed by later work, was that the really small particles of
coal (<19μm diameter)are a major source of adhesion between
the larger coal particles and the particles of inorganic

sulphur and ash. It was concluded that dry separation was
viable if these ultra-fines were removed prior to separation.
These findings were supported by a major DOE-funded project
carried out at Oak Ridge National Laboratory in association with
Union Carbide (ref.115). In this report are summarized
the results of over one thousand laboratory tests carried
out on a variety of Eastern U.S. coals. In these tests a compar-
son was made of the various leading techniques for applying
H.G.M.S. to industrial coals. The powdered coals were supplied
to the matrix in one batch of tests as a gravity-fed dry
powder (along with a hot air stream), in another as an aqueous
slurry, in another as a dry fluidized bed, and, lastly,
as a dry powder pumped upwards through the matrix against gravity.
These tests appeared to demonstrate the superiority of dry
gravity feeding over other competing schemes, provided that
ultra-fines are removed from the feed. An extension of this work
carried out recently at Sala Magnetics Inc. involved dry gravity
feeding of Eastern U.S. coals at a rate of $1th^{-1}$ on a pilot
scale continuous carousel separator of the type illustrated in
Fig.34. This pilot scale plant was shown to be very efficient
and was demonstrated publicly to leading technologists from
the coal industry at Cambridge, Ma. on April 24,1980.

It is worth noting briefly three further points on work in
this area. Y.A.Lui and co-workers report (ref.116)   successful
separation of a variety of industrial dry coals using a novel
fluidized bed containing a magnetic matrix. Here, it is found
that the above-mentioned ultra fines need not be removed from
the feed prior to separation. However, the rather long
residence times and rather small loading factors on the matrix
raise some doubt as to the economic viability of large-scale
processing by this technique. Secondly, a method of treatment
of dry coals using conventional magnetic separation equipment
has been reported by Kindig (ref.117).   This is the Magnex
process in which finely powdered coal is treated with iron
carbonyl, a chemical vapour which enhances the magnetic
susceptibility of the inorganic sulphur and ash compounds
but leaves the coal unaffected. The dry powdered coal is then
subjected to conventional magnetic separation systems. Furthermore
E C Hise and co-workers (ref.94) have recently carried out

successful separations on gravity-fed dry powdered coals at
Oak Ridge using a novel open gradient separator. The open
gradient principle is essentially that described in §4.5 for
the Frantz isodynamic separator in the free-fall mode. There
is no particle capture. Instead the magnetic field gradients act
as a beam splitter, deflecting the product and the mags to
separate geographical locations in the device. No detailed
report has been published of the finished design of this system
at the time of writing of this review but it would appear that
the field gradients will be generated by a superconducting pair
of Helmholtz coils operating in opposition. Finally, Dijkhuis
and Kerkdijk have recently reported informally (ref.13) on inorganic
sulphur from powdered coal in a carrier stream of liquid nitrogen.
This technique offers several advantages over other schemes.
First, this wet separation technique is enhanced by the flow
viscosity of the liquid(roughly four times smaller than water).
Secondly, the susceptibility of the inorganic sulphur is much
higher than at room temperatures, in accordance with Curie's
law (ref.18). Finally, the liquid nitrogen can be used in an
integral system which includes cryogenic grinding of the coals.
Parker (ref.118) has shown recently that this technique also
facilitates the use of matrices comprising fibres of rare earth
metals such as Gd,Dy,Tb which have exceptionally large values
of magnetization at cryogenic temperatures.

6.4 Other developing applications

In addition to coal and to kaolin, a variety of other
mineral ores can be upgraded using either cyclic or continuous
Kolm Marston systems of the sort described in §4.6.

. Until recently, low grade iron ores, which are abundant
throughout the world, have received little attention in
regard to exploitation, particularly in the United States(ref.119).
These ores are, in the main, oxidised, taconites in
which the magnetite (Fe_3O_4) fractions have been oxidised
to the more weakly magnetic haematite (Fe_2O_3) and in
which fractions comprising iron silicates and siderite
have been converted to the more weakly magnetic goethite,
a hydrated iron oxide (ref.120). Moreover, such ores, normally,
must be ground to very fine sizes to liberate the iron content.
A pilot scale version of the continuous Carousel separator

of Fig.34 was used around 1974 by Kelland and Maxwell in a series of laboratory and field tests at Nasevauk, Minn., on these types of ores (ref.119). The feasibility of their exploitation was demonstrated clearly. A more recent detailed study by Hopstock (ref.120) has confirmed this but has shown also the desirability of HGMS techniques being used in conjunction with selective flocculation and froth flotation.

Other important new areas of investigation in the field of minerals beneficiation include the reduction of copper in molybdenum concentrates, of iron oxide in glass sands, barites and ceramic clays, as well as purification of phosphates and various refractory materials. The carousel continuous HGMS system is particularly effective in the concentration of various rare earth minerals and in the superconcentration of iron ores, producing less than 2% silicon in the latter at competitive levels of cost. It has been shown possible to remove tourmaline from tin (cassiterite) ores. It has also been shown possible to produce preconcentrates for flotation and other processes. Mineral ores which may be so treated include those of uranium and lithium.

Dense media separation in the coal industry using concentrated slurries of magnetite has already been mentioned in connection with wet drum technology. Advances in the cleaning of smaller particles of coal has brought about the necessity of feeding streams of coal and magnetite directly to separator systems. Here, wet drum systems have not been satisfactory in terms of the level of recovery of magnetite fines. Kelland and Maxwell (ref.121) have succeeded with this type of feed, in obtaining 99% recovery of magnetite with less than 5% coal entrapment in a single stage Koln-Marston system with a specially designed matrix.

A number of small-scale investigations of medical and biological applications of HGMS must be mentioned. The work of Bitton and Mitchell (ref.103) in virus removal and in the recovery of single cell protein has already been referred to. Melville and co-workers (refs.122,123) have demonstrated the feasibility of separating red blood cells from whole blood. Timbrell (ref.124) has used HGMS techniques to capture respirable asbestos fibres from air streams. Little published work has appeared in the area of HGMS applied to particulates carried in air streams. However, an impressive experimental study of HGMS applied to iron oxide forms in steel mill operations has been carried out by Gooding and co-workers (ref.125) at the Research Triangle Institute in North Carolina. Some of their bench-scale measurements of the successful recovery of BO furnace dust in an air stream with a flow velocity of 8.4 ms^{-1} and a filter pressure drop of 1.7 kPa are shown in Fig.72, for a range of applied fields and dust particles diameters.

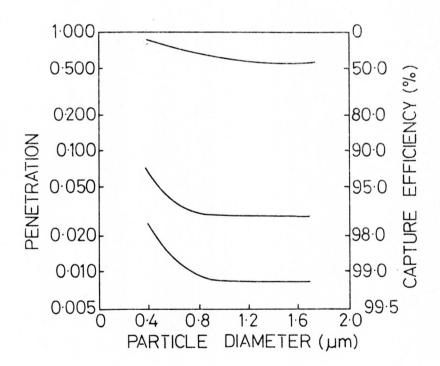

Fig.72. Collection efficiency of BOF dust with a velocity of 8.4ms^{-1} and a filter pressure drop of 1.7 kPa (after Gooding et al ref.125)

ACKNOWLEDGEMENTS

The authors gratefully acknowledge the generous help received in the preparation of this manuscript from Dr O Endogen Gurmen of Stearns Magnetics Inc, Dr I S Wells of Boxmag-Rapid Ltd., Dr K-H Unkelbach of KHD Industrienlagen A.G, Mr Thomas D Wellington of S.G.Frantz-Company Inc., Mr Henry Ruff and Dr Bengt Anden of Sala International and Mr Brian Jepson of E.C.L.P and Co.Ltd. The authors also thank Dr R Gerber for many useful discussions and are indebted to Mr M K Wong for generous communication of graphic material.

APPENDIX 1

a) Magnetic Potential of a Cylindrical Fibre

Consider an infinite cylindrical fibre (Fig.36), of radius a, placed transversely in an external magnetic field, H_o, sufficient to magnetize the fibre to saturation. By considering the known solutions of Laplace's equation in cylindrical geometry (ref.18), it may be seen that the magnetostatic potential of the fibre may be expressed, in cylindrical polar coordiantes, as

$$\emptyset = Cr\cos\theta + (D/r)\cos\theta$$

where C and D are constants to be determined from the boundary conditions. In general terms, magnetostatic potential and magnetic field $\underline{H}$ are related by the expression

$$\underline{H} = -\nabla\emptyset \qquad (A1)$$

Inside the cylinder the field is uniform. Consequently, the potential, $\emptyset_1$, within the fibre is given by

$$\emptyset_1 = -H_i r\cos\theta$$

Obviously, a non-zero value of the constant D would result in an infinite potential at the origin of Fig.36.

Outside the cylinder, as $r \to \infty$, the field becomes uniform and equal to the external magnetic field, H_o, giving

$$\emptyset_2 = -H_o r\cos\theta + (D/r)\cos\theta \qquad (A2)$$

the second term on the RHS of eqn.(A2) becoming negligible as

$r \to \infty$. The constant D may be evaluated in terms of the system parameters by consideration of the continuity of the normal component of the magnetic induction, $\underline{B}$, and the tangential component of the magnetic field, $\underline{H}$, at the surface of the fibre. The first of these conditions gives

$$-\mu_o \frac{\partial \emptyset_1}{\partial r} + M_s \cos\theta = -\mu_2 \frac{\partial \emptyset_2}{\partial r} \tag{A3}$$

where μ_o μ_2 are the permeabilities of free space and of the fluid, respectively. The second condition gives

$$-(1/r) \frac{\partial \emptyset_1}{\partial \theta} = -(1/r) \frac{\partial \emptyset_2}{\partial \theta} \tag{A4}$$

leading to values of the internal field H_i and of the constant, D, of

$$H_i = (M - 2\mu_o H_o)/(\mu_o + \mu_o) \tag{A5}$$

and $\quad D = \{H_o a \ (\mu_o - \mu_2) + M_s a^2\} \ /(\mu_2 + \mu_o).$ $\qquad$ (A6)

For all normal fluids, $\mu_2 \simeq \mu_o$, producing the result

$$\emptyset_1 = (M_s/2\mu_o - H_o) r \cos\theta \tag{A7}$$
$$\emptyset_2 = -H_o r \cos\theta + (M_s a^2/2\mu_o r) \cos\theta \tag{A8}$$

From eqns.(A1) and (A8) it follows that the radial component of the local field in the on-axis ($\theta = o$) direction is

$$H_r(\theta = o) = -(H_o + \frac{M_s a^2}{2\mu r_o^2}) \tag{A9}$$

At the surface of the fibre, the radial component of the field gradient is therefore

$$\frac{\partial H_r}{\partial r} (\theta = o) = M_s/\mu_o a$$

Using eqn.(A8), the function $\nabla(\underline{H}^2)$ may be evaluated, in cylindrical polar coordinates, for the specific case of the above-described infinite cylindrical fibre. In cylindrical polar coordinates, $\nabla \emptyset = \hat{r} \frac{\partial \emptyset}{\partial r} + \hat{\theta}(1/r) \frac{\partial \emptyset}{\partial \theta} + \hat{z} \frac{\partial \emptyset}{\partial z}$

where $r, \theta, \underline{z}$ are unit vectors along the polar axes. Here,

$\frac{\partial \emptyset}{\partial \underline{z}}$ may be ignored since H_o is orthogonal to the z-axis (the fibre axis). Therefore, since $\underline{H} = -\nabla\emptyset$(ref.18)

$$\underline{H} = \left(\frac{\partial\emptyset_2}{\partial r} \right)^2 \left[(1/r^2)\frac{\partial\emptyset_2}{\partial\theta} \right]^2 \tag{A10}$$

Substituting for $\emptyset_2$ from eqn.(A9) into eqn.(A10) gives

$$\underline{H}^2 = H_o^2 \cos \theta + 2(H_oA/r^2)\cos^2\theta + (A^2/r^4)\cos^2\theta$$

$$+ H_o^2\sin^2\theta - 2(H_oA/r^2)\sin^2\theta + (A^2/r^4)\sin^2\theta \tag{A11}$$

where $A = M_s a^2/2\mu_o$. $\tag{A13}$

Thus $\underline{H}^2 = H_o^2 + (2H_oA/r^2)\cos 2\theta + A^2/r^4$

From eqn. (4) and eqn. (A13) the radial and azimuthal components of the magnetic tractive force exerted by the magnetized fibre on a non-ferromagnetic particle of volume V_p and of net volume susceptibility $\chi(=\chi_p - \chi_m)$ are respectively

$$F_{M,r} = - \frac{2\mu_o\chi V_pA}{r^3} \left(\frac{A}{r^2} + H_o\cos 2\theta \right) \tag{A14}$$

$$F_{M,\theta} = \frac{2\mu_o\chi V_pAH_o\sin 2\theta}{\theta r^3} \tag{A15}$$

b) Fluid Potential of a Cylindrical Fibre

The distribution of fluid flow potential around the fibre of Fig.36 is obtained in a similar manner to that used above to calculate the magnetic potential. In conditions of inviscid flow, Lamb's potential, $\emptyset_F$, for fluid flow is

$$\emptyset_F = - V_orcos(\theta-\gamma) + (D_F/r)\cos(\theta-\gamma), \tag{A16}$$

where, at large values of r, γ is the angle between the direction of flow and the $\theta=0$ direction and where V_o is the magnitude of the fluid velocity. The radial component of the fluid velocity must obviously be zero at the surface of the fibre ($r=a$) and so, from eqn.(A16)

From eqn. (4) and eqn.(A13) the radial and azimuthal components of the magnetic tractive force exerted by the magnetized fibre on a non-ferromagnetic particle of Volume v_p and of net volume susceptibility $/(=/_p-/_m)$ are respectively

$$F_{M,r} = - \frac{2\mu_0/v_p A}{r^3} \; (\frac{A}{r^2} + H_0 \cos 2\theta) \tag{A14}$$

$$F_{M,\theta} + - \frac{2\mu_0/V_p AH_0 \sin 2\theta}{r^3} \tag{A15}$$

c) Fluid Potential of a Cylindrical Fibre

The distribution of fluid flow potential around the fibre of Fig.36 is obtained in a similar manner to that used above to calculate the magnetic potential. In conditions of inviscid flow, Lamb's potential, $\emptyset_F$, for fluid flow is

$$\emptyset_F = - V_0 r\cos(\theta-\gamma) + (D_F/r)\cos(\theta-\gamma), \tag{A16}$$

where, at large values of r, is the angle between the direction of flow and the $\theta=0$ direction and where V_0 is the magnitude of the fluid velocity. The radial component of the fibre velocity must obviously be zero at the surface of the fibre ($r=a$) and so, from eqn. (A16)

$$0 = -V_0 \cos(\theta-\gamma) - (D_F/a^2)\cos(\theta-\gamma).$$

Thus $D_F = -a^2 V$ leading to the result that

$$\emptyset_F = -V_0 r\cos(\theta-\gamma) - V_0 (a^2/r)\cos(\theta-\gamma). \tag{A17}$$

Since $V_0(r) = - \frac{\partial \emptyset_F}{\partial r}$ and $V_0(\theta) = (1/r) \frac{\partial \emptyset_F}{\partial \theta}$

the components of the fluid velocity may be expressed as

$$V(r)/a = (V_0/a)(1-1/r_a^2)\cos(\theta-\gamma) \tag{A18}$$

$$V(\theta)/a = (V_0/a)(1+1/r_a^2)\sin(\theta-\gamma) \tag{A19}$$

In the L-configuration, $\gamma = 0$. In the T-configuration $\gamma = \pi/2$.

Appendix 2

Some manufacturers of high
intensity magnetic separation
equipment

(a) UNITED STATES

Company	Activities
Sala Magnetics Inc. 169 Bent Street Cambridge Mass. 0214	Conventional high intensity magnetic separators.Large-scale HGMS systems (including carousel designs). Test facilities.
Stearns Magnetics Inc., 6001 S.General Avenue Cudany Wisconsin 53 UG	Complete range of conventional separation systems: test facilities
Pacific Electric Motor Co. 1009 6th Avenue Oakland California 94621.	Conventional (water-cooled, iron-bound solenoid) HGMS systems.
Magnetic Corporation of America 179 Bear Hill Road Waltham Mass.02154.	Conventional and superconducting systems; test facilities.
Aquafine Corporation 157 Darien Highway Brunswick Georgia 31520	Test facilities for industry.
Carpco Research and Engineering Inc., Jacksonville Florida	Conventional high intensity wet and dry separation systems.
Intermagnetics General Co. PO Box 566 Guilderland. NY 612084	Superconducting systems.
Eriez Magnetics Inc., 23rd & Astbury Road Erie Pa.16512.	Permanent magnet separation systems.

(b) UNITED KINGDOM

Boxmag Rapid Ltd. Chester St. Aston Birmingham B6 4AJ	Complete range of conventional separation systems: test facilities.
Jones Separators Ltd Connor Downs Hoyle Cornwall.	Jones separators

International Research &
Development Co.Ltd.
Fossway
Newcastle upon Tyne NE62YD

Large conventional test systems;
superconducting magnet specialists.

Cryogenic Consultants
Metrostore Building
231 The Vale
London W3

Specialists in superconducting
quadropole and open gradient
separator systems.

(c) FEDERAL REPUBLIC OF GERMANY

K.H.D.
Instrianlangen AG
5 Koln 91
Postfach 910404

Complete range of conventional
separation systems: Jones
separators; test facilities.

Kraftwerk Union AG
Hammerbacherst 12 + 14
Postfach 3220
8520 Erlangen

Electromagnetic filters for large
scale nuclear power plants.

Appendix 3

The following instructions provide the programme procedure to
be undertaken for the solution of eqn.(40). In this example the
L-configuration is selected, with $\quad$ = 0. The method of solution
involves a fourth-order Runge-Kutta subroutine provided on an
optional Math/Utilities Chip of a Texas Instruments TI59
Calculator. The procedure summarised below is dervied from
the User Instructions MU-18 described in the T.I.Publication
No.1014984 - 10. This particular programme is designed to
solve an equation of the type $\frac{d\theta}{dr_a} = f(r_a, \theta)$. The Runge-Kutta
method consists of finding approximate solutions at
particular points r , r , r , ... where the differences
between successive r values $(r_{a(n+1)} - r_{an})$ is constant. The
value of this difference is selected by the user with smaller
values yielding more accurate results. In this programme the
difference in adjacent r_a coordinates is chosen by the
number of divisions chosen (see STEP No.86 below), large
numbers corresponding to smaller differences in r_a. It is also
necessary to enter a subroutine defining $\frac{dr_a}{dt}$ (STEP No.71)
and to enter r_{ai}, θ_i (STEP Nos.82 to 86). Finally, it is
necessary (STEP N o.88) to select a new value of r_a for which
a solution is desired.

STEP	ENTER	PRESS	STEP	ENTER	PRESS
1		(	46		x
2		(	47		1/x
3		11	48		)
4		RCL	49		x
5		11	50		RCL
6		x	51		10
7		1/x	52		Sin
8		)	53		-
9		x	54		RCL
10		RCL	55		00
11		10	56		x
12		Cos	57		(
13		+	58		RCL
14		RCL	59		11
15		00	60		x
16		x	61		2
17		(	62		)
18		RCL	63		sin
19		01	64		÷
20		÷	65		RCL
21		RCL	66		10
22		11	67		y
23		y	68		3
24		5	69		)
25		+	70		1/x
26		(	71		INV
27		RCL	72		SBR
28		11	73		2ND
29		x	74		Pgm
30		2	75		18
31		)	76	$\lvert V_m/V_o \rvert$	
32		Cos	77		STO
33		÷	78		00
34		RCL	79	K	STO
35		10	80		STO
36		3	81		01
37		)	82	$\lvert$initialθ(rd$\rvert$	
38		)	83		A
39		÷	84	$\lvert$initial $r_a \rvert$	
40		(	85		B
41		(	86	$\lvert$No.of divisions$\rvert$	
42		1	87		D
43		+	88	$\lvert$new $r_a \rvert$	
44		RCL	89		2ND
45		11	90		E'

→

The program prints out a new value of θ corresponding to the new value of r_a. Steps 82 to 85 are then repeated with these revised values of r_a and θ and this procedure continues until a final value of r_a is arrived at.

References

1.   British Patent in 1792 to William Fullarton.

2.   F Langguth, "Handbuch der Elektrochemie Electro-
      magnetische Aufbereitung"(Halley A.S., 1903).

3.   D Korda, "La Separation Electromagnetique et
      Electrostatique des Mineraux"(L'Eclairage
      Electrique, Paris, 1905).

4.   C G Gunther, "Electro-magnetic Ore Separation"(McGraw-
      Hill, New York, 1909).

5.   R S Dean and C W Davis, "Magnetic Separation of Ores",
      (Bureau of Mines Bulletin 425, U.S.Dept. of the
      Interior, 1941).

6.   E J Pryor, "Mineral Processing" (Elsevier, New York,
      1965), Chap.19.

7.   A F Taggart, " Elements of Ore Dressing" (John Wiley,
      New York, 1951).

8.   R Mitchell, G.Bitton and J A Oberteuffer, "Separation
      and Purification Methods" (Marcel Dekker, New York,
      1976) 267-303.

9.   J A Oberteuffer, "Magnetic Separation : A Review of
      Principles, Devices and Applications", IEEE, Trans.
      Magn, MAG-10, (1974), 223-238.

10.   D R Kelland, E.Maxwell and J A Oberteuffer, "High
      Gradient Magnetic Separation: An Industrial Application
      of Magnetism", Proceedings of NATO Institute on Super-
      conducting Machines and Devices: Large System
      Applications, Francis Bitter National Magnet Laboratory,
      Cambridge, Mass., 1974.

11.   H H Kolm, J A Oberteuffer and D R Kelland, "High
      Gradient Magnetic Separation", Sci.Am., 223, (1975)
      46-54.

12.   M R Parker, "The Physics of Magnetic Separation", Contemp.
      Phys., 18, (1977), 279-306.

13.   J I Dijkhuis and C B W Kerkdijk, "Upgrading of Coal using
      Cryogenic HGMS" (submitted to IEEE, Trans.on Magnetics).

14.   R R Birss and M R Parker, "Magnetic Separation of Dia-
      magnetic Particles", IEEE, Trans. Magn., MAG-15,
      (1979/ 1523-5.

15.   E Schloemann, "Eddy Current Separation Methods", Progress
      in Filtration and Separation, 1, (1979) 29-82.

16.   F D DeVaney, "New Developments in the Magnetic Concentration
      of Iron Ores", 5th Int.Minerals Processing Cong.,
      London (1960) 745-749.

17.   R R Birss, R Gerber, M R Parker and T J Sheerer,
       "Theory and Performance of Axial Magnetic Filters in
       Laminer Flow Conditions", IEEE Trans. Magn.,
       MAG-14, (1978), 389-391.

18.   B I Bleaney and B Bleaney, "Electricity and Magnetism",
       (1st.Edn.), (Clarendon, 1959).

19.   A Aharoni, "Traction Force on Paramagnetic Particles in
       Magnetic Separators", IEEE.Trans.Magn., MAG-12,(1976),
       236-235.

20.   M K Wong, (private communication)

21.   M R Parker, "Domain Wall Capture of Magnetic Particles",
       Univ.of Salford Seminar, November (1978).

22.   E Laurila, "On the Fields and Permanent Magnets in
       Magnetic Pulleys and Separators", Acta Polytechnica
       Scandinavia (1962), 312.

23.   R R Birss, M R Parker and M K Wong, "Modeling of Fields
       in Magnetic Drum Separators", MAG-15,(1979) 1305-09.

24.   P J Kihlstedt and B Skold, "Concentration of Magnetite
       Ores with Dry Magnetic Separators of the Martsell-Sala
       Type", 5th Int.Minerals Processing Cong., London,
       (1963), 691-704.

25.   G Grondal, U.S.Patent No.691,262(1902).

26.   J Suleski, "New Magnets and Tank Designs for wet
       Magnetic Drum Separators," World Mining, (1972), 60-61.

27.   R E Crockett, U.S.Patent N o.2,003,430 (1935).

28.   C Q Payne, U.S.Patent, No.901,368 (1908).

29.   S G Frantz, U.S.Patent No.2,074,085 (1937).

30.   Thomas D Wellington (S.G.Frantz Company Inc) (private
       communication).

31.   G H Jones, British Patent No. 768,451 (1955).

32.   G H Jones, "Wet Magnetic Separator for Feebly Magnetic
       Minerals", Proc.7th Int.Minerals Processing Cong.
       (1964) 717-732.

33.   W H Simons and R P Treat, "Results on Particle
       Trajectories in a Lattice of Magnetized Fibres in
       Gradient Magnetic Separation", Proc.Conf.on Ind.
       Applications of Mag.Sepn. IEEE Publn. No.78CH1447-2MAG
       (1979)145-149.

34.   I Isenstein, "Magnetic Traction Force in an HGMS with
       an Ordered Array of Wires, I & II, IEEE.Trans.Magn.
       MAG-14, (1978) 1148-1157.

35.   S Uchiyama and K Hayashi, "Analytical Theory of
        Magnetic Particle Capture Process and Capture Radius
        in High Gradient Magnetic Separation", Proc.Int.Conf.on
        Ind. Applications of Mag.SepnIEEE Publication
        No.78CH1447 -2MAG (1979), 169-173.

36.   R P Walker, MSc.Dissertation (University of Salford),
        1980.

37.   K Hayashi and S Uchiyama, "On Particle Trajectory and
        Capture Efficiency Around Many Wires", IEEE Trans.
        on Magn.MAG-I6 (1980). 827-829.

38.   H G Heitmann, "Studies of the Application of Electro-
        Magnetic Filters in Power Plants", Proc.Int.Conf.on
        Ind.Applications of Mag.Sepn. IEEE Publn.No.78CH1447-
        2MAG (1979) 115-120.

39.   L Dolle, P Grandcollot, J.Chenouard, R Darras and
        P Beslu, "Extraction of Insoluble Corrosion Products
        from the Water Circuits of Nuclear Power Stations by
        Electromagnetic Filtration", Proc.of 2nd World
        Filtration Cong,, London, (1979) 617-628.

40.   J McAndrew, "Calibration of a Frantz Isodynamic
        Separator and its Application to Mineral Separation",
        Proc.Aus.I.M.M., (1957) 59-73.

41.   W M Kester, J W Leonard and E B Wilson, "Reduction of
        Sulphur from Steam Coal by Magnetic Methods",
        Mining Congress Journal, (1970) 70-73.

42.   H H Kolm and D B Montgomery, Proc.Conf.on High Magnetic
        Fields and their Applications", Nottingham (1969)
        (unpublished).

43.   H H Kolm, U.S. Patent No. 3567026 (1971).

44.   R R Birss, R Gerber and M R Parker, "Analysis of
        Matrix Systems in High Intensity Magnetic Separation",
        Filtration and Separation, (1977), 339-342.

45.   R R Birss, R Gerber and M R Parker, "Theory and
        Design of Axially Ordered Filters for High Intensity
        Magnetic Separation", IEEE Trans.Magn. MAG-12,
        (1976) 892-894.

46.   C P Bean, Bull.Am.Phys.Soc., 16, (1971) 340.

47.   J H P Watson, "Magnetic Filtration", J.App.Phys., 44,
        (1973), 4209-4213.

48.   J E Nesset and G A Finch, "A Static Model of High
        Gradient Magnetic Separation Based on Forces within the
        Fluid Boundary Layer", Proc.Int.Conf.on Ind.Application
        of Mag.Sepn., IEEE Publn.No.78CH1447-2 MAG, (1979),
        188-196.

49.   C J Clarkson, "The Modelling of High Gradient Magnetic
      Separation and its Application to Cold Rolling Mill
      Lubricants", PhD. Thesis, M.I.T. (1970).

50.   A Probert, Undergraduate Student Report, Dept. of
      Physics, University of Salford (1980).

51.   P G Marsden, U.S.Patent No. 3627678 (1971).

52.   R R Birss, J M Dyson, M R Parker and T J Sheerer,
      "Fluid Impedance Characteristics of Filters for HGMS",
      Proc.Ind.World Filtration Cong.(1979), 471-477.

53,   R R Birss, R Gerber and M R Parker, British Patent.
      No.1562 941 (1977).

54.   I N Bronshtein and K A Semendyayev, "A Guide Book to
      Mathematics", (Springer-Verlag, N.Y.Inc.) (1973),
      160-164.

55.   R Gerber (private communication).

56.   F E Luborsky and B J Drummond, "High Gradient Magnetic
      Separation : Theory vs. Experiment", IEEE. Trans.
      Magn., MAG-11, (1975), 1696-1700.

57.   C Cowen, F J Friedlaender,and R Jaluria, "Single Wire
      Model of High Gradient Magnetic Separation Processes 1*"
      IEEE.Trans.Magn.,MAG-12,(1976)

58.   R R Birss and M R Parker, "Domain Wall Capture of
      Magnetic Particles", J.Magn. & Magn.Mater., 15-18,
      (1980), 1567-1568.

59.   W F Lawson, Jr., W H Simons and R P Treat, "The Dynamics
      of a Particle Attracted by a Magnetized Wire", J.App.
      Phys., 48, (1977) 3213-3217.

60.   S Uchiyama, S Kondo, M Takayasu and I Eguchi,
      "Performance of Parallel Stream Type Magnetic Filter
      for HGMS", IEEE Trans.Magn., MAG-12, (1976) 895-897.

61.   R R Birss, R Gerber, M R Parker and T J Sheerer,
      "Laminar Flow Model of Particle Capture of Axial
      Magnetic Filters", IEEE Trans.Magn., MAG-14 (1978)
      1165-1169.

62.   R R Birss, R Gerber, M R Parker, and T J Sheerer,
      "Characteristic Properties of Axial Filters for High
      Gradient Magnetic Separation", Proc.Conf.on Ind..
      Applications of Magn.Sepn., IEEE Publn.No.78 CH 1447 -
      2 MAG (1979) 174.

63.   T J Sheerer, PhD.Thesis, University of Salford (1981)

64.   F E Luborsky and B J Drummond, "Buildup of Particles on
      Fibers in a High Field-High Gradient Separator",
      IEEE Trans.Magn. MAG-12 (1976) 463-465.

65. H Schlichting, "Boundary Layer Theory", McGraw-Hill, New York, (1960) Chap.X11.

66. J H P Watson, "Superconducting Magnetic Separation at Moderate Reynolds Number", Proc.XV Int.Cong. of Refrigeration, Venice, (1979) A1/2-23.

67. J H P Watson, "Particle Retention in High-Intensity Magnetic Separation", Proc.Symp.on Deposition and Filtration of Particles from Gas and Liquids, Loughborough University (1978).

68. R B Bird, W E Stewart and E N Lightfoot, "Transport Phenomena", John Wiley, New York, (1960) 59.

69. J H P Watson, "HGMS at Moderate Reynolds Number", IEEE Trans.Magn., MAG-15, (1979) 1538.

70. L Milne-Thomson, "Theoretical Hydrodynamics"(5th Edition), McMillan, London, (1968).

71. S Uchiyama, S Kurinobu, M Kumazawa and M Takayasu, "Magnetic Particle Buildup Process in Parallel Stream Type HGMS Filter", IEEE Trans.Magn.,MAG-14, (1977), 1490-1492.

72. C Cowen, F J Friedlaender and R Jaluria, "Single Wire Model of High Gradient Magnetic Separation Processes", IEEE Trans.Magn., MAG-11, (1975) 1600-1603; ibid IEEE Trans.Magn., MAG-12, (1976) 898-900.

73. F J Friedlaender, M Takayasu, J B Rettig and C P Kentzer, "Studies of Single Wire Parallel Stream HGMS", IEEE Trans.Magn., MAG-14 (1978), 404-406.

74. F J Friedlaender and M Takayasu", Video Recording of Particle Trajectories and Buildup of Single Wires on High Gradient Magnetic Separation" Proc.on Conf. on Ind.App. of Magn. Sepn., IEEE Publ.No.78CH/467-2MAG, (1979) 154-158.

75. W H McNeese, P C Wankat, F J Friedlaender, T Nakano and M Takayasu, "Viscosity Effects in Single Wire HGMS Studies" IEEE Trans.on Magn., MAG-15, (1979) 1520-1522.

76. F J Friedlaender, M Takayasu, T Nakano and W H McNeese, "Diamagnetic Capture in Single Wire HGMS", IEEE.Trans. Magn., MAG-15, (1979) 1526-1528.

77. W H McNeese, P C Wankat and F J Friedlaender, "Viscosity Effects in Single Wire HGMS Studies II", IEEE. Trans. Magn., MAG-16, (1980) 843-845.

78. M K Wong, Undergraduate Student Project, University of Salford, (1978).

79. J E Nesset and J A Finch, "Fine Particles Processing", 2, P.Somasunderan, ed., New York; AIME,(1980) 1217-1241.

300

80. J E Nesset, I Todd, M Hollingworth and J A Finch,
    "A Loading Equation for High Gradient Magnetic
    Separation", IEEE.Trans.Magn., MAG-16 (1980).833-835.

81. H K Collan, J Jantunen, M Kokkala and A Ritvos, "Inversion
    of the Breakthrough Curve of a High Gradient Magnetic
    Filter: Theory and Experiment", Proc.Int.Conf. on Inds.
    App. of Mag.Sepn., IEEE, Publn.No.78 CH 1447-2MAG (1979),
    175-187.

82. R R Birss, M R Parker and T J Sheerer, "Statistics of
    Particle Capture in HGMS", IEEE Trans.Magn.MAG-16, (1980)
    830-832.

83. C J Clarkson and D R Kelland, "Theory and Experimental
    Verification of a Model for High Gradient Magnetic
    Separation", IEEE, Trans.Magn.MAG-14, (1978), 97-103.

84. G Dobby, J Nesset and J A Finch, Mineral Recovery by High
    Gradient Magnetic Separation", Canadian Metallurgical
    Quarterly, 18, (1979), 293-301.

85. I Y Akoto, "Mathematical Modelling of High Gradient Magnetic
    Separation Devices", IEEE Trans.Magn.MAG-13,(1977)
    1486-1489.

86. J H P Watson, "Approximate Solutions of the Magnetic
    Separation Equations", IEEE Trans.Magn., MAG-14,
    (1978) 240-245.

87. S Uchiyama and S Kurinobu, "Steady-State Solution of Filter
    Equations of High Gradient Magnetic Separation", Proc.Int.
    Conf.on Ind.Applications, of Mag.Sepn., IEEE. Publn.No.
    78CH 1447-2MAG (1979) 166-168.

88. H K Collan. M Kokkala and A Ritvos, "Analysis of Magnetic
    Filter Experiments with Polydisperse Particle Suspensions",
    IEEE Trans.Magn., MAG-15, (1979), 1529-1531.

89. R R Oder and C R Price, "HGMS: Mathematical Modelling
    of Commercial Practice", A.I.P. Conference Proc.(1975),
    641-643.

90. R R Birss, M R Parker and T J Sheerer, "A Radioactive
    Tracer Technique for the Evaluation of Magnetic Filter
    Performance", IEEE Trans.Magn., MAG-16 (1980),
    836-839.

91. H Kolm: unpublished work in the use of superconducting
    quadrupole magnets for the beneficiation of molybdenum
    ores.

92. E Maxwell, "Magnetic Separation - The Prospects for
    Superconductivity", Cryogenics (1975) 179-184.

93. H E Cohen and J A Good,"Magnetic Separation of Mineral
    Slurries by a Cryogenic Magnet System", Filtration and
    Separation, (1977), 346-348.

94. E C Hise, A S Holman and F J Friedlaender, "Development of High Gradient and Open Gradient Magnetic Separation of Dry Fine Coal", Digests of the 1981 Intermag Conference, Grenoble, (to be published).

95. J H P Watson and D Hocking, "The Beneficiation of Clay using a Superconducting Magnetic Separator", IEEE Trans. Magn., MAG-17, (1975), 1588-1590.

96. J A Oberteuffer and I Wechsler, PRecent Advances in High Gradient Magnetic Separation", Proc.of A.I.M.E. Conf. on Fine Particle Processing, Las Vegas (1980).

97. J A Oberteuffer, I Wechsler, P G Marston and M J McNallan, "High Gradient Magnetic Filtration of Steel Mill Process and Waste Waters", IEEE Trans.Magn., MAG-11, (1975), 1591-1593.

98. K Takino, T Takayoshi and T Schickiri, "Operation of a Production-Scale High Gradient Magnetic Filter for Treatment of Gas Scrubber Water from a Steel Mill Vacuum Degassing Process", Proc.Int.Conf. on Ind.Application of Magn.Sepn., IEEE Publn. No.78CH-1447-2MAG(1979), 137-141.

99. J Yano and I Eguchi, "Application of High Gradient Magnetic Separation for Water Treatment in Steel Industry", Proc. Int.Conf.on Ind.Applications of Mag.Sepn., IEEE.Pubn. No.78CH1447 2 Mag, (1979) 134-136.

100. M Troy (Principal Investigator),"Study of Magnetic Filtration Applications to the Primary and Secondary Systems of PWR Plants," Electric Power Research Institute (EPRI), Palo Alto, 1978, Publn.Nl. EPRI-NP-514TPS-76-665.

101. H H Elliott, J H Holloway and D G Abbot, "The Potential Uses of High Gradient Magnetic Filtration for High-Temperature Water Purification in Boiling Water Reactors", Proc.Int.Conf.on Ind.Applications of Mag.Sepn., IEEE. Publ.No. 78CH1447-2MAG, (1979) 127-134.

102. C DeLatour, "Magnetic Separation in Water Pollution Control", IEEE Trans.Magn.,MAG-9,(1973) 314-316.

103. C B itton and R Mitchell, "The Removal of Escherichia Coli - Bacteriophage T_7 by Magnetic Filtration",Water Research, 8, (1974) 549-551.

104. L Petrakis and P F Ahner, "High Gradient Magnetic Separation on Water Effluents", IEEE. Trans.Mag.MAG-14(1978)481-493.

105. A Z Yurovsky and I D Remesnikov, "Thermomagnetic Method of Concentrating and Desulphurizing Coal", Coke and Chemistry, (1958).

106. R D Harris, "Reducing the Sulphur Content of Steam Coal by Removing Fine Iron Pyrite at the Power Station". Paper presented at the Fall Meeting Society of Mining Engineers of AIME, October 7-9, Phoenix, Arizona, (1965).

107. S Trinidade and H H Kolm, "Magnetic Desulphurization of Coal", IEEE.Trans.Magn., MAG-9, (1973)310-311.

108. E Maxwell and D R Kelland, "High Gradient Magnetic Separation in Coal Desulphurization", IEEE Trans.Magn. MAG-14, (1978) 482-487.

109. R E Hucko, "DOE Research in High Gradient Magnetic Separation Technology", Proc.Int.Conf. on Ind. Applications of Mag.Sepn., IEEE, Publn.No. 78CH1447-2MAG, (1979), 77-82.

110. E.P.R.I. 389, Final Report, 2, February (1976).

111. Y A Liu and C J Lin, "Assessment of Sulphur and Ash Removal from Coals by Magnetic Separation", IEEE. Trans.Magn., MAG-12, (1976), 583-550.

112. E Maxwell, E.P.R.I. Report AF-508.

113. I S Jacobs, L M Levinson and H R Hart, Jr., "Magnetic and Mossbauer Spectroscopic Characterization of Coal", J.App.Phys., 49, (1978) 1775-1780.

114. F E Luborsky, Bureau of Mines SRD-77-147, (Contract No.HO366008), September 19, (1977).

115. E C Hise, I Wechsler and J M Doulin, "Separation of Dry Crushed Coals by High-Gradient Magnetic Separation, DOE Report (Contract No.W-7405-eng-26) October(1979).

116. Y A Liu, C J Lin and D M Eisenberg, "A Novel Fluidized Bed Dry Magnetic Separation Process with Applications to Coal Beneficiation", Digests of the 1978 Intermag Conf.(Florence).

117. J K Kindig, "The Magnet Process: Review and Current Status", Proc.Int.Conf.on Ind.Applications of Mag. Sepn. IEEE, Publn.No.78CH1447-2MAG, (1979) 79-104.

118. M R Parker (to be published)

119. D R Kelland and E Maxwell, "Oxidized Taconite Beneficiation by Continuous High Gradient Magnetic Separation", IEEE. Trans.Magn.MAG-11, (1975),1582-1584.

120. D M Hopstock, "Wet High-Intensity Magnetic Beneficiation of Oxidised Taconites", Proc.Int.Conf.on Ind. Applications of Magn.Sepn. IEEE, Publn.No.78CH1447-2MAG, (1979)55-61.

121. D R Kelland and E Maxwell, "Improved Magnetite Recovery in Coal Cleaning by HGMS", IEEE.Trans.Mag., MAG-14, (1978) 401-403.

122. D Melville, F Paul, and S Roath, "High Gradient Magnetic Separation of Red Cells from Whole Blood", IEEE Trans. Magn., MAG-11, (1975), 1701-1703.

123. D Melville, "Magnets for Separating Blood", Spectrum, (1975), 5-6.

124. V Timbrell, "Magnetic Separator of Respirable Asbestos
      Fibres", Proc.2nd.Conf.on Adv.in Mag.Mat. and their
      Applications", London (1976) 89-91.

125. C H Gooding, T W Signman, L K Monteith and D C Drehmel,
      "Application and Modelling of High Gradient Magnetic
      Filtration in a Particulate Gas System", IEEE.Trans.
      Magn., MAG-14, (1978) 407-409.

PROGRESS IN FILTRATION AND SEPARATION 1
Edited by R.J. Wakeman

CONTENTS

306